VEHICULAR 2018

The Seventh International Conference on Advances in Vehicular Systems, Technologies and Applications

June 24 - 28, 2018

Venice, Italy

VEHICULAR 2018 Editors

Manabu Tsukada, University of Tokyo, Japan

Masaaki Sato, Keio University, Japan

Markus Ullmann, Federal Office for Information Security / University of Applied Sciences Bonn-Rhine-Sieg, Germany

Printed from e-media with permission by:

Curran Associates, Inc.
57 Morehouse Lane
Red Hook, NY 12571

Some format issues inherent in the e-media version may also appear in this print version.

Printed by Curran Associates, Inc. (2018)

International Academy, Research, and Industry Association (IARIA)
412 Derby Way
Wilmington, DE 19810

Phone: (408) 893-6407
Fax: (408) 527-6351

petre@iaria.org

Additional copies of this publication are available from:

Curran Associates, Inc.
57 Morehouse Lane
Red Hook, NY 12571 USA
Phone: 845-758-0400
Fax: 845-758-2633
Email: curran@proceedings.com
Web: www.proceedings.com

Table of Contents

VEHICULAR 2018

Forward

The Seventh International Conference on Advances in Vehicular Systems, Technologies and Applications (VEHICULAR 2018), held between June 24, 2018 and June 28, 2018 in Venice, Italy, continued a series of events considering the state-of-the-art technologies for information dissemination in vehicle-to-vehicle and vehicle-to-infrastructure and focusing on advances in vehicular systems, technologies and applications.

Mobility brought new dimensions to communication and networking systems, making possible new applications and services in vehicular systems. Wireless networking and communication between vehicles and with infrastructure have specific characteristics from other conventional wireless networking systems and applications (rapidly-changing topology, specific road direction of vehicle movements, etc.). These led to specific constraints and optimizations techniques; for example, power efficiency is not as important for vehicle communications as it is for traditional ad hoc networking. Additionally, vehicle applications demand strict communications performance requirements that are not present in conventional wireless networks. Services can range from time-critical safety services, traffic management, to infotainment and local advertising services. They are introducing critical and subliminal information. Subliminally delivered information, unobtrusive techniques for driver's state detection, and mitigation or regulation interfaces enlarge the spectrum of challenges in vehicular systems.

The conference had the following tracks:

- Fundamentals on communication and networking
- Experiments and challenges
- Security and evaluation
- Unmanned vehicles
- Cooperative Intelligent Transportation Systems (CITS)

We take here the opportunity to warmly thank all the members of the VEHICULAR 2018 technical program committee, as well as all the reviewers. The creation of such a high quality conference program would not have been possible without their involvement. We also kindly thank all the authors who dedicated their time and effort to contribute to VEHICULAR 2018. We truly believe that, thanks to all these efforts, the final conference program consisted of top quality contributions.

We also gratefully thank the members of the VEHICULAR 2018 organizing committee for their help in handling the logistics and for their work that made this professional meeting a success.

We hope that VEHICULAR 2018 was a successful international forum for the exchange of ideas and results between academia and industry and to promote further progress in the field of vehicular systems, technologies and applications. We also hope that Venice, Italy provided a pleasant environment during the conference and everyone saved some time to enjoy the unique charm of the city.

VEHICULAR 2018 Chairs

VEHICULAR Steering Committee

Markus Ullmann, Federal Office for Information Security / University of Applied Sciences Bonn-Rhine-Sieg, Germany
Carlos T. Calafate, Universitat Politècnica de València
Éric Renault, Institut Mines-Télécom | Télécom SudParis, France
Samy El-Tawab, James Madison University, USA
Khalil El-Khatib, University of Ontario Institute of Technology - Oshawa, Canada
Manabu Tsukada, University of Tokyo, Japan

VEHICULAR Industry/Research Advisory Committee

Clément Zinoune, Renault, France
Michelle Wetterwald, HeNetBot, France
Yi Ding, US Army RDECOM-TARDEC, USA
William Whyte, Security Innovation, USA

VEHICULAR 2018

Committee

VEHICULAR Steering Committee

Markus Ullmann, Federal Office for Information Security / University of Applied Sciences Bonn-Rhine-Sieg, Germany
Carlos T. Calafate, Universitat Politècnica de València
Éric Renault, Institut Mines-Télécom | Télécom SudParis, France
Samy El-Tawab, James Madison University, USA
Khalil El-Khatib, University of Ontario Institute of Technology - Oshawa, Canada
Manabu Tsukada, University of Tokyo, Japan

VEHICULAR Industry/Research Advisory Committee

Clément Zinoune, Renault, France
Michelle Wetterwald, HeNetBot, France
Yi Ding, US Army RDECOM-TARDEC, USA
William Whyte, Security Innovation, USA

VEHICULAR 2018 Technical Program Committee

Ali Abedi, University of Maine, USA
Eric J. Addeo, DeVry University, USA
Sufyan T. Faraj Al-Janabi, University of Anbar, Ramadi, Iraq
Michele Albano, CISTER | INESC/TEC | Polytechnic Institute of Porto, Portugal
Alexandre Armand, Renault, France
Kamran Arshad, Ajman University, United Arab Emirates
Dalila B. Megherbi, University of Massachusetts, USA
Eduard Babulak, Fort Hays State University, USA
Andrea Baiocchi, University of Roma "Sapienza", Italy
Stylianos Basagiannis, United Technologies Research Centre, USA
Marcel Baunach, Graz University of Technology | Institute for Technical Informatics, Austria
Luca Bedogni, University of Bologna, Italy
Rahim (Ray) Benekohal, University of Illinois, USA
Chafika Benzaid, University of Sciences and Technology Houari Boumediene (USTHB), Algeria
Luis Bernardo, Universidade NOVA de Lisboa, Portugal
Neila Bhouri, IFSTTAR / COSYS / GRETTIA, France
Abdelmadjid Bouabdallah, Université de Technologie de Compiègne, France
Christos Bouras, University of Patras | Computer Technology Institute & Press «Diophantus», Greece
Salah Bourennane, Ecole Centrale Marseille, France
Alexander Brummer, University of Erlangen-Nürnberg, Germany
Lin Cai, Illinois institute of Technology, USA
Carlos T. Calafate, Universitat Politècnica de València

Roberto Caldelli, CNIT - National Interuniversity Consortium for Telecommunications | MICC - Media Integration and Communication Center, Universita' degli Studi Firenze, Italy
Maria Calderon, University Carlos III of Madrid, Spain
Juan-Carlos Cano, Universidad Politécnica de Valencia, Spain
Rodrigo Capobianco Guido, São Paulo State University at São José do Rio Preto, Brazil
Juan Carlos Ruiz, Universitat Politecnica de Valencia, Spain
Sandra Céspedes, Universidad de Chile, Santiago, Chile
Amitava Chatterjee, Jadavpur University, Kolkata, India
Claude Chaudet, Webster University Geneva, Switzerland
Olfa Chebbi, Higher Institute of Management of Tunis, Tunisia
Jiajia Chen, KTH Royal Institute of Technology, Sweden
Mu-Song Chen, Da-Yeh University, Taiwan
Dong Ho Cho, School of Electrical Eng., KAIST, Korea
Gihwan Cho, Chonbuk University, Korea
Woong Cho, Jungwon University, South Korea
Theofilos Chrysikos, University of Patras, Greece
Pierpaolo Cincilla, Technological Research Institute SystemX (IRT-SystemX), France
Yousef-Awwad Daraghmi, Palestine Technical University-Kadoorie, Palestine
Klaus David, University of Kassel, Germany
David de Andrés, Universitat Politècnica de València, Spain
Carl James Debono, University of Malta, Malta
Sergio Di Martino, University of Naples Federico II, Italy
Omar Dib, Technological Research Institute IRT SystemX, France
Yi Ding, US Army RDECOM-TARDEC, USA
Sanjay Dorle, G. H. Raisoni College of Engineering, Nagpur, India
Mariagrazia Dotoli, Politecnico di Bari, Italy
David Eckhoff, TUMCREATE, Singapore
Ghaïs El Zein, Institut d'Electronique et de Télécommunications de Rennes (IETR) | Institut National des Sciences Appliquées (INSA), France
Safwan El Assad, University of Nantes, France
Mohamed El Kamili, University Sidi Mohammed Ben Abdellah of Fez, Morocco
Khalil El-Khatib, University of Ontario Institute of Technology - Oshawa, Canada
Samy El-Tawab, James Madison University, USA
Marcos Fagundes Caetano, University of Brasília, Brazil
Camille Fayollas, ICS-IRIT | University Toulouse 3 Paul Sabatier, France
Gustavo Fernandez, Austrian Institute of Technology, Austria
Gianluigi Ferrari, University of Parma, Italy
Attilio Fiandrotti, Politecnico di Torino, Italy
Miguel Franklin de Castro, Federal University of Ceará, Brazil
Małgorzata Gajewska, Gdansk University of Technology, Poland
Sławomir Gajewski, Gdansk University of Technology, Poland
David Gallegos, Applus+ IDIADA, Spain
Peter Gaspar, MTA SZTAKI, Hungary
Hao Ge, Northwestern University, USA
Reinhard German, UniversityofErlangen-Nuremberg, Germany
Thanassis Giannetsos, University of Surrey, UK
Apostolos Gkamas, University Ecclesiastical Academy of Vella of Ioannina, Greece
Javier Gozalvez, Universidad Miguel Hernandez de Elche, Spain

Luigi Alfredo Grieco, Politecnico di Bari, Italy
Stefanos Gritzalis, University of the Aegean, Greece
Yu Gu, Hefei University of Technology, China
Mesut Güneş, Institute for Intelligent Cooperating Systems | Otto-von-Guericke-University Magdeburg, Germany
Florian Hagenauer, Paderborn University, Germany
Biao Han, National University of Defense Technology, Changsha, China
Petr Hanáček, Brno University of Technology, Czech Republic
Hong Hande, National University of Singapore, Singapore
Morteza Hashemi, Ohio State University, USA
Hiroyuki Hatano, Utsunomiya University, Japan
Ivan W. H. Ho, The Hong Kong Polytechnic University, Hong Kong
Javier Ibanez-Guzman, Renault S.A., France
Khalil ibrahimi, Ibntofail University, Kenitra, Morocco
Hocine Imine, IFSTTAR/LEPSIS, France
Mohammad Reza Jabbarpour, Islamic Azad University, Tehran-North Branch, Iran
Terje Jensen, Telenor, Norway
Han-You Jeong, Pusan National University, Korea
Yiming Ji, University of South Carolina Beaufort, USA
Magnus Jonsson, Halmstad University, Sweden
Yasin Kabalci, Omer Halisdemir University,Turkey
Arnaud Kaiser, Institut de Recherche Technologique SystemX (IRT SystemX), France
M. A. S. Kamal, Monash University Malaysia, Malaysia
Frank Kargl, Ulm University, Germany
Sokratis K. Katsikas, University of Piraeus, Greece
Abdelmajid Khelil, Landshut University, Germany
Wooseong Kim, Gachon University, Korea
Xiangjie Kong, Dalian University of Technology, China
Jerzy Konorski, Gdansk University of Technology, Poland
Dimitrios Koukopoulos, University of Patras, Greece
Zdzislaw Kowalczuk, Gdansk University of Technology, Poland
Milan Krbálek, Czech Technical University in Prague, Czech Republic
Ajith Kumar P. R., Nokia, Bangalore, India
Anh Le Tuan, Insight - NUI Galway, Ireland
Fedor Lehocki, Slovak University of Technology in Bratislava, Slovak Republic
Christian Lehsing, Harvard Medical School, USA & Technical University of Munich, Germany
Pierre Leone, University of Geneva, Switzerland
Marco Listanti, University of Roma "La Sapienza", Italy
Lianggui Liu, Zhejiang Sci-Tech University, China
Miao Liu, IBM Research, USA
Yali (Tracy) Liu, AT&T labs, Inc., USA
Seng Loke, Deakin University, Melbourne, Australia
Xuanwen Luo, Sandvik Mining, USA
Barbara M. Masini, CNR - IEIIT, Italy
Leandros Maglaras, De Montfort University, UK
Abdelhamid Mammeri, University of Ottawa, Canada
Zoubir Mammeri, IRIT - Paul Sabatier University, France
Chetan Belagal Math, Eindhoven University of Technology, Netherlands

Natarajan Meghanathan, Jackson State University, USA
Rashid Mehmood, King Abdul Aziz University, Jeddah, Saudi Arabia
João Mendes-Moreira, LIAAD-INESC TEC | University of Porto, Portugal
Lyudmila Mihaylova, University of Sheffield, UK
Vicente Milanes, RENAULT SAS, France
Steffen Moser, Ulm University, Germany
Nils Muellner, Malardalen University, Sweden
Saeid Nahavandi, Institute for Intelligent Systems Research and Innovation (IISRI) - Deakin University, Australia
Mort Naraghi-Pour, Louisiana State University, USA
Jose E. Naranjo, INSIA | Technical University of Madrid, Spain
António J. R. Neves, University of Aveiro, Portugal
Tae (Tom) Oh, Rochester Institute of Technology, USA
Arnaldo Oliveira, Universidade de Aveiro, Portugal
Tomas Olovsson, Chalmers University of Technology, Sweden
Rachid Outbib, Aix-Marseille University, France
Markos Papageorgiou, Technical University of Crete, Greece
Al-Sakib Khan Pathan, Southeast University, Bangladesh
Xiaohong Peng, Aston University, UK
Fernando Pereñíguez García, Politécnica Universidad Católica San Antonio Murcia (UCAM), Spain
Hamid R. Rabiee, Sharif University of Technology, Iran
Ali Rafiei, University of Technology Sydney, Australia
Jacek Rak, Gdansk University of Technology, Poland
Hesham Rakha, Virginia Tech Transportation Institute, USA
Mubashir Husain Rehmani, COMSATS Institute of Information Technology, Pakistan
Éric Renault, Institut Mines-Télécom | Télécom SudParis, France
M. Elena Renda, IIT - CNR - Pisa, Italy
Jean-Pierre Richard, Centrale Lille - French "Grande Ecole", France
Martin Ring, Hochschule Karlsruhe, Germany
Justin P. Rohrer, Naval Postgraduate School, USA
Claudio Roncoli, Aalto University, Finland
Javier Rubio-Loyola, CINVESTAV, Mexico
Larry Rudolph, TWO SIGMA, LP, USA
Marcel Rumez, University of Applied Sciences, Karlsruhe, Germany
José Santa Lozano, University of Murcia, Spain
Panagiotis Sarigiannidis, University of Western Macedonia, Greece
Christoph Schmittner, AIT Austrian Institute of Technology GmbH, Austria
Erwin Schoitsch, AIT Austrian Institute of Technology GmbH, Austria
Miguel Sepulcre, Universidad Miguel Hernandez de Elche, Spain
Zhengguo Sheng, University of Sussex, UK
Shinichi Shiraishi, Toyota InfoTechnology Center, USA
Mohammad Shojafar, University of Rome, Italy
Dana Simian, Lucian Blaga University of Sibiu, Romania
Dimitrios N. Skoutas, University of the Aegean, Greece
Mujdat Soyturk, Marmara University, Turkey
Vasco N. G. J. Soares, Affiliation: Instituto de Telecomunicações / Instituto Politécnico de Castelo Branco, Portugal
Peter Steenkiste, Carnegie Mellon University, USA

Reinhard Stolle, BMW Group, Germany
Thomas Strubbe, Federal Office for Information Security (BSI), Germany
Nary Subramanian, The University of Texas at Tyler, USA
Young-Joo Suh, Postech (Pohang University of Science & Technology), Korea
Muhammad Tahir, Lahore University of Management Sciences (LUMS), Pakistan
Necmi Taspinar, Erciyes University, Turkey
Daxin Tian, Beihang University, China
Tammam Tillo, Libera Università di Bolzano-Bozen, Italy
Angelo Trotta, University of Bologna, Italy
Eirini Eleni Tsiropoulou, University of Maryland, College Park, USA
Manabu Tsukada, University of Tokyo, Japan
Bugra Turan, Koc University, Istanbul, Turkey
Ion Turcanu, Sapienza University of Rome, Italy
Piotr Tyczka, ITTI Sp. z o.o., Poznań, Poland
Markus Ullmann, Federal Office for Information Security / University of Applied Sciences Bonn-Rhine-Sieg, Germany
Rens W. van der Heijden, Ulm University, Germany
John Vardakas, Iquadrat Informatica, Barcelona, Spain
Quoc-Tuan Vien, Middlesex University, UK
Massimo Villari, Universita' di Messina, Italy
Ljubo Vlacic, Griffith School of Engineering, Australia
Lingfeng Wang, University of Wisconsin-Milwaukee, USA
You-Chiun Wang, National Sun Yat-sen University, Taiwan
Shuangqing Wei, Louisiana State University, USA
Andre Weimerskirch, Lear Corporation, USA
Michelle Wetterwald, HeNetBot, France
William Whyte, Security Innovation, USA
Safwan Wshah, PARC - a Xerox Company, USA
Fan Wu, Tuskegee University, USA
Pei Xiao, Institute for Communication Systems (ICS) | University of Surrey, UK
Ramin Yahyapour, Gesellschaft für wissenschaftliche Datenverarbeitung mbH Göttingen (GWDG), Germany
Fan Ye, Stony Brook University, USA
Shingchern D. You, National Taipei University of Technology, Taiwan
Chuan Yue, Colorado School of Mines, USA
S.M. Salim Zabir, National Institute of Technology | Tsuruoka College, Japan
David Zage, Intel Corporation, USA
Sherali Zeadally, University of Kentucky, USA
Degan Zhang, Tianjin University of Technology, China
Liang Zhang, Magna Electronics, Brampton Ontario, Canada
Zhiyi Zhou, Northwestern University, USA
Clément Zinoune, Renault, France

Copyright Information

Efficient V2X Data Dissemination in Cluster-Based Vehicular Networks

Yongyue Shi, Xiao-Hong Peng
School of Engineering & Applied Science
Aston University
Birmingham, UK
Email: {shiy9, x-h.peng}@aston.ac.uk

Guangwei Bai
Department of Computer Science & Technology
Nanjing Tech University
Nanjing, China
Email: bai@njtech.edu.cn

Abstract— **Vehicular communications and networking technologies provide essential support for data services across a Vehicular Ad-hoc Network (VANET) and play a key role in the Intelligent Transport System (ITS). In this paper, we introduce a cluster-based two-way data service model that promotes efficient cooperation between Vehicle-to-Vehicle (V2V) and Vehicle-to-Infrastructure (V2I) communications, or namely V2X, to improve service performance for vehicles and the network. Our results show that the cluster-based model can significantly outperform the conventional non-cluster schemes, in terms of service successful ratio, network throughput and energy efficiency.**

Keywords*— *cluster; V2X communications; VANET; energy efficiency.

I. INTRODUCTION

With the rapidly increasing number of vehicles and complex road networks, traffic congestion, car accidents and large amount of energy consumption are among the main challenges in the development of smart mobility as part of the Intelligent Transportation System (ITS) [1]. To address these problems and ensure road safety and traffic efficiency, it is vital to make traffic information (e.g., speed and vehicle density) and environmental information (e.g., weather and road condition) timely available for road users and network operators.

The Vehicular Ad-hoc Network (VANET) is an extended version of the Mobile Ad-hoc Network (MANET) and intended for improving driving safety and efficiency through both vehicle-to-vehicle (V2V) and Vehicle-to-Infrastructure (V2I) communications. V2V and V2I can be operated cooperatively as V2X, making the VANET play a better role in ITS in a complex traffic environment.

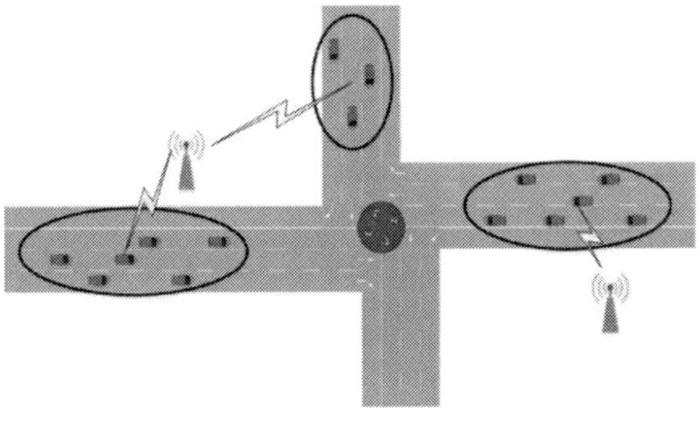

Figure 1. A VANET model with clusters

This paper proposes a V2X-based service system where the clustering technique is applied to improve transmission and energy efficiencies by significantly reducing V2I connections. A cluster is a group of vehicles within the transmission range of each other, as shown in Fig. 1 where cluster heads exchange data with RSU via V2I while the other cluster members communicate with cluster heads via V2V. A data service model with cooperative V2X transmission via clustering is also introduced, for effectively uploading the local information to the database and downloading the required service data from RSUs.

The remaining of the paper is organized as follows. Related work is discussed in Section II. Section III presents the clustering algorithm and applies it in the proposed data service model. Section IV explains the simulation results produced by OMNET++, SUMO and MATLAB software tools. Finally, Section V concludes the paper.

II. RELATED WORK

The idea of combining V2I and V2V has been applied in many works on VANET. In [2], Noori et al. explore the combination of various forms of communication techniques, e.g., cellular network, Wi-Fi and ZigBee for VANETs. In [3], a roadside unit (RSU) plays a vital part to provide services and make scheduling arrangements using a simple network coding in a V2X approach. This approach may cost more energy to complete the service and does not consider the packet loss and associated latency caused by the failed services. In [4], multiple RSUs are involved in broadcasting data periodically to vehicles via V2I and forwarded to vehicles via V2V if they are not inside the transmission range. This model requires efficient handover mechanisms to ensure stable and in-time data services between the vehicles concerned.

The Dedicated Short-Range Communications (DSRC) technology refers to a suite of standards of Wireless Access in Vehicular Environments (WAVE) [5] and supports both V2V and V2I communications. Vehicles equipped with sensors can collect local traffic and environment information and exchange it for the similar information of other regions (place of interest) with RSUs. A RSU acts as an interface between vehicles and the vehicular network to provide vehicles the service information requested and pass on the collected information to other part of the network. The high mobility and density of vehicles presents a big challenge in V2X communications, which causes congestion in service delivery in this environment. In addition, moving vehicles will keep exchanging information and this will cost a significant amount energy for continuous data sensing, transmission and processing, especially for V2I as it needs to cover longer distances than V2V.

The Lowest-ID clustering algorithm is a basic method to select cluster head, which uses the unique vehicle ID numbers as the selection standard [6]. This algorithm works stably in most MANETs but may not always be suitable for VANET due to higher velocity and more restricted routes for vehicles. The AMAC (Adaptable Mobility-Aware Clustering) algorithm [7] mainly considers the destination as the key factor in forming clusters to improve the stability of clusters and extend the cluster's lifetime. However, the destination may not always be collected from navigation systems as drivers do not always use them for the known routes. A three-layer cluster head selection algorithm based on the interest preferences of vehicle passengers is proposed for multimedia services in a VANET [8]. This scheme is inefficient when the requirements in operations differ too much.

Based on the discussion of V2X related work, a more efficient service delivery method is introduced in this paper by utilizing clusters and minimizing channel congestion caused by excessive V2I transmission in conventional service models. We will show that the cluster model outperforms the non-cluster model at both service and energy efficiency levels.

III. SERVICE MODEL THROUGH CLUSTERING

In MANETs, moving nodes can be divided into different sizes of clusters, such as using the "combined weight" algorithm to select cluster heads [9]. The selection takes the current position, number of neighbours, mobility, and battery power of nodes into consideration. In VANETs, vehicles' mobility is more limited by the road type, traffic signs and other traffic factors. Therefore, the elements involved in forming clusters in a VANET need to be adjusted accordingly.

A. Cluster Head Selection

There are three types of nodes (vehicles) in a VANET: Free Node (FN), Cluster Head (CH), Cluster Member (CM). The clustering algorithm considers the one-hop neighbours of each node and the cluster size is decided by cluster head's communication range. CH is responsible for collecting data and service requests from CMs, uploading current driving information (e.g., traffic is normal or congested), and requesting services from the RSUs. This paper defines a new weighting metric for selecting the CH, considering the factors, such as position, velocity, connectivity and driving behaviour of the vehicles involved.

The position of each node is obtained from GPS (Global Positioning System) data. The average distance, P_i, between CH and CM should be as short as possible, which is given by

$$P_i = \frac{1}{n}\sum_{j=1}^{n}\sqrt{(x_j - x_i)^2 + (y_j - y_i)^2} \quad (1)$$

where n is the number of neighbours of node n_i, x and y are coordinate values of two involved nodes.

The velocity of CH, V_i, is defined to be the difference between the velocity of a candidate node v_i and the average velocity for the current traffic flow, and given by:

$$V_i = \left| v_i - \frac{1}{n}\sum_{j=1}^{n} v_j \right| \quad (2)$$

where v_j is the velocity of the j-th neighbour of the candidate node.

The connectivity of the candidate node is reflected by the number of its neighbours, N_i. The ideal connectivity is denoted as σ, which represents the maximum number of neighbouring nodes within one hop without causing traffic congestion, and is given as:

$$\sigma = 2R_t \times 133 \times n_l / 1000 \quad (3)$$

where R_t is the transmission range, n_l is the number of lanes. The constant value 133 represents the highest possible density (vehicles/(lane·km) [10]. The actual connectivity, C_i, is to measure how close the N_i is to the ideal value σ, i.e.:

$$C_i = |N_i - \sigma| \quad (4)$$

The last factor is the acceleration of the vehicle, a_i, to reflect the driving behaviour D_i by showing how stable a vehicle is when running along the road, i.e.:

$$D_i = |a_i| \quad (5)$$

The weighting matrix is formed by combining the four factors discussed above which are considered equally important. After the normalization of the four measurements, as shown below,

$$\text{a) } P_i' = \frac{P_i}{P_{\max i}}, \text{ b) } V_i' = \frac{V_i}{V_{\max}}, \text{ c) } C_i' = \frac{C_i}{\sigma}, \text{ d) } D_i' = \frac{D_i}{D_{\max i}} \quad (6)$$

the weighting matrix, , is defined as

$$W_i = P_i' + V_i' + C_i' + D_i' \quad (7)$$

where P_{max} is the distance between the i-th vehicle and the farthest vehicle from it, V_{max} is the speed limitation by traffic rules that a vehicle can reach in the flow, D_{max} is the maximum absolute value of acceleration the vehicle can reach when it is running. A smaller W_i indicates the higher suitability of the candidate for the CH.

When a vehicle detects itself as a free node (FN), it sends a vehicle information packet to its neighbours and enables them to calculate its weight value *Wi* based on (7) which is the basis of CH selection: the vehicle with the smallest Wi value becomes the CH. If a vehicle generates a *Wi* that is smaller than a weight threshold, it will send a claim message with its weight *Wi* to the neighbours to announce its suitability for CH. Other nodes will compare the received *Wj* with their own weight and send claim messages to argue if theirs have a smaller weight than *Wi*. Otherwise, it will become the CH after a threshold window time and declare its identity as CH of its neighbours. This process takes place at either a fixed or varied interval(s) depending on traffic conditions given.

The service model that we have developed utilizes cluster-based V2X communications. In this model, vehicles are grouped into clusters for information exchange between vehicles and RSUs. CHs are selected to gather and aggregate information collected by CMs and disseminate service packets to CMs via V2V. V2I transmissions take place only between CHs and RSUs via V2I directly, including uploading information to the server via RSU and downloading service data from RSU by CHs, as shown in Fig. 2.

The cluster-based service model has transferred most of the data delivery from long-range V2I to short-range V2V. In this way, both transmission collision in the vehicle-RSU links and energy consumption can be reduced. The database server shown in Fig. 2 stores service information including the traffic and environmental information such as the velocity of current traffic flow, real-time density of vehicles, weather conditions and road status, which is updated periodically.

This service system follows the standards of IEEE 802.11p and IEEE 1609 family [5], which specifies 7 channels of 10 MHz each including one control channel and 6 service channels. The control channel is used for exchanging control messages and safety information, while service channels are used for delivering service information packets.

B. Service Delivery

Vehicles within the same cluster may gather similar information, especially the weather and road conditions. In addition, different vehicles may request information for the same regions. Therefore, CH integrates the collected information before forwarding it to RSU. The aggregated data at CH will be less than what it has been collected, so the transmission efficiency in the V2I links can be improved. Upon receiving the information from CH, RSU updates the database and generates the service packets requested by vehicles. Service packets are then sent via V2I to CH which will redistribute them to CMs via V2V.

Each RSU maintains its own database to store the recent service information collected from different CHs within its coverage. RSUs in different areas will periodically exchange and update information between them. In this case, vehicles in one area can learn the information about a larger range of areas ahead. The information service helps drivers to choose the best routes to reach their destinations and avoid congestion and accidents. They can also be aware of the travelling time they will spend.

The RSU is up to 8-15 meters high [11] and the distance between a RSU and vehicles is much farther than the distance between vehicles themselves, thus V2I requires higher transmitting power than V2V to deliver data. The transmitting power for V2V mainly depends on the distance between CH and the farthest CM from CH and the maximum transmission distance (d^*) in this case is mainly based on the number of vehicles in a cluster. Denote the distance between two vehicles as $d_{i,j}$, then:

$$d^* = \max_{i \in n} \max_{j \in n} \{d_{i,j}\} \tag{8}$$

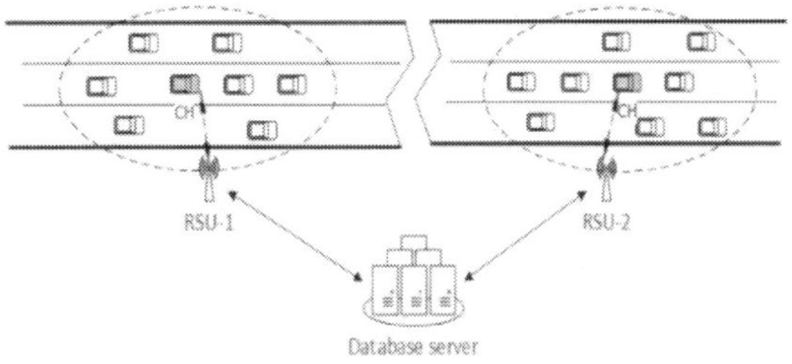

Figure 2. Cluster-based service model

Given the receiver sensitivity P_r, the required transmitting power of the i-th transmission by a vehicle, P_{ti}, is given by

$$P_{ti} = P_r \cdot L_{pi} \tag{9}$$

where L_{pi} is the path loss of this transmission link and represented by (assuming the free-space scenario)

$$L_{pi} = \left(\frac{4\pi d_i}{\lambda}\right)^2 G_t G_r \tag{10}$$

where d_i is the link distance, λ is the wavelength of the signal transmitted, G_t and G_r are transmitting and receiving antenna gains.

For cluster-based service delivery, the total transmitting power, P_{tc}, can be calculated as:

$$P_{tc} = \sum_{i=1}^{n-1} P_{V2V-i} + P_{V2I} \tag{11}$$

where n is the number of vehicles in a cluster, P_{V2V} is the transmitting power for V2V communications, defined as P_{ti} in (9), and P_{V2I} is the transmitting power for V2I communications.

For service delivery without clusters, the total transmitting power P_t, is simply the sum of individual vehicle transmission power all in the V2I mode, i.e.:

$$P_t = \sum_{i=1}^{n} P_{V2I-i} \tag{12}$$

C. Performance Evaluation

In this paper, the following four metrics are applied to evaluate the performance of the proposed system.

- Service ratio (γ). It is the ratio of the number of successful delivered requests n_s to the total number of requested services n. This is a vital metric to evaluate the effectiveness of the V2X system. This performance metric is given by:

$$\gamma = \frac{n_s}{n} \tag{13}$$

- Average service delay (τ). It is defined as the average duration from a vehicle submitting a service request to it finally receiving the service packets, which is expressed by:

$$\tau = \frac{\sum_{i=1}^{n_s} t_{si} + n_{us} \cdot t_p}{n_s} \quad (14)$$

where t_{si} is the time duration of the i-th successful service transmission, n_{us} is the number of unsuccessful service requests, and t_p is the waiting time a vehicle spends for the service which is not delivered.

- Throughput (η). It is a widely applied metric to evaluate the transmission efficiency of a system. It is defined as the average size of data successfully delivered over a time unit.

$$\eta = \frac{p_s}{T} \quad (15)$$

where p_s is the total size of delivered service packets, T is the total transmission time.

- Energy Consumption (E_C). It is measured as an average amount of energy (Joule) consumed for transmitting one bit of data, or called energy per bit. Given transmitting power P_t and throughput η, the energy consumption is given by

$$E_C = \frac{P_t}{\eta} \quad (16)$$

IV. SIMULATION AND RESULTS ANALYSIS

A. Simulation Setup

The traffic scenarios and communications models are simulated using SUMO [12] and OMNET++ [13]. SUMO is a powerful traffic simulator and supports multiple road topologies and vehicle attributes. It can cooperate with other network simulators via its Traffic Control Interface (TraCI) modules. OMNET++ is an extensible, modular, and component-based C++ simulation framework, supporting various types of network simulation developments.

We built a one-way straight road with three lanes on SUMO, in which vehicles in each lane are running as a flow and the related service model is shown in Fig. 2. According to the Highway Code [14], the safe stopping distances are related to the driving speed. Considering the transmission range of V2V, which is usually 300 metres, the number of vehicles in a cluster on motorways is related to the flow speed as well. Based on the safe stopping distance, we define six scenarios in simulation for the flow speed of 32, 48, 64, 80, 96, 112 km/h, respectively. The relationship between the vehicle number and flow speed is shown in Fig. 3.

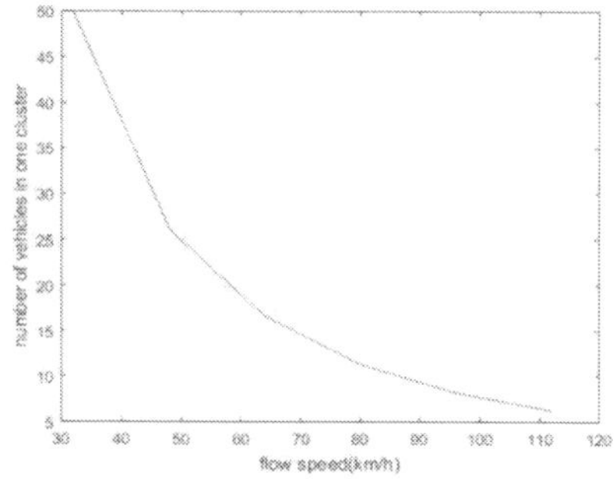

Figure 3. Service ratio under different flow speeds

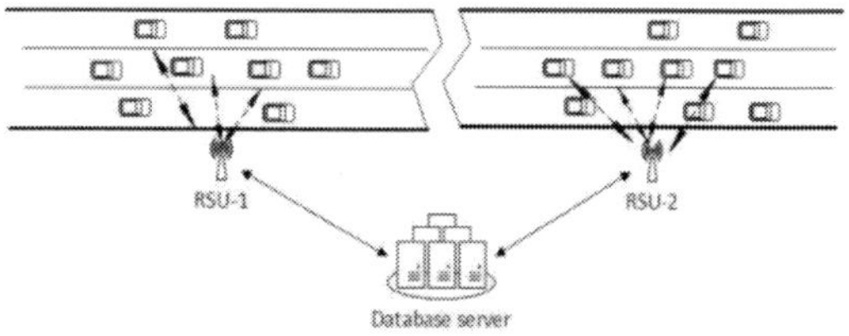

Figure 4. Non-cluster service model

TABLE I. SIMULATION PARAMETERS

Parameters	Value
Frequency band	5.850-5.925 GHz
Channel bandwidth	10 MHz
Receive power sensitivity	-89dBm
Propagation model	Free space model
Data rate	6Mbps, 12Mbps
Number of requests	20-25
Data size	1000 bits
Number of lanes	3
Simulation time	300s

The transmission model is configured based on the IEEE 802.11p and IEEE 1609 Family. Table I gives the parameters of the physical and MAC (Media Access Control) layers of the vehicular communication system and Table II specifies the

TABLE II. TRANSMISSION POWER IN V2V AND V2I

Flow speed *(km/h)*	32	48	64	80	96	112
V2V *(mW)*	0.802	1.020	0.899	0.867	0.925	0.711
V2I *(mW)*	2.885	2.898	2.890	2.841	2.878	2.821

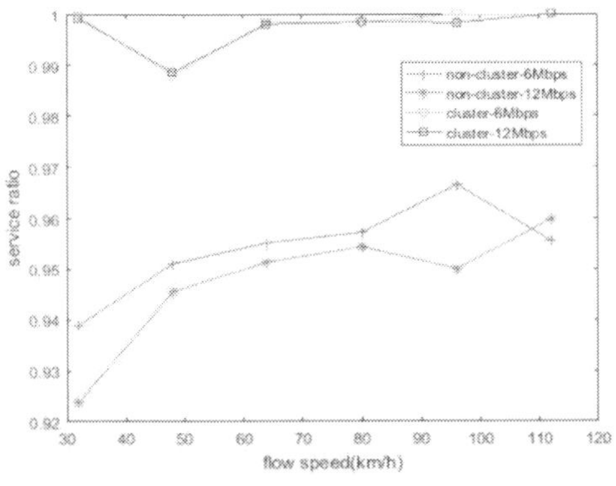

Figure 5. Service ratio under different flow speeds

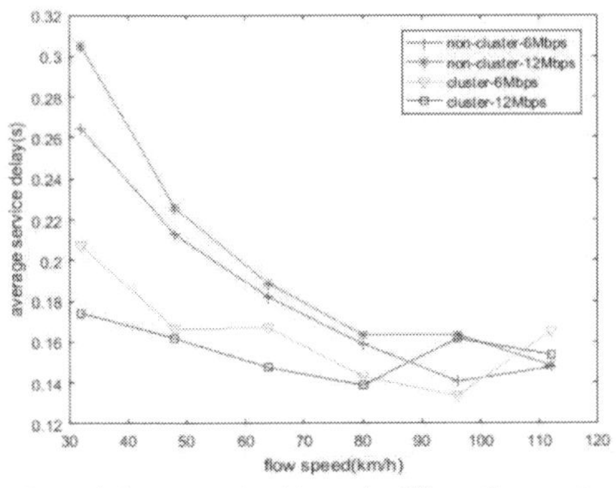

Figure 6. Average service delay under different flow speeds

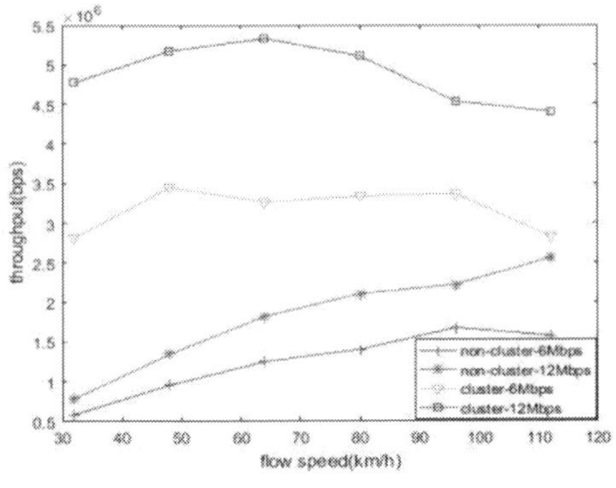

Figure 7. Throughput under different flow speeds

transmission power in different modes (V2V and V2I), which are adopted in simulations.

For the purpose of performance comparison with the proposed service model, we have also simulated the non-cluster model, as shown in Fig. 4 where the same number of vehicles and vehicle velocity are set in each scenario. Once the vehicles enter the transmission range of the RSU, they communicate with RSU directly via V2I. The two models are evaluated for the same set of performances, featuring the service ratio, average service delay, throughput and energy consumption.

B. Results analysis

Fig. 5 shows different service ratios (or successful rate of service delivery) of both Cluster-Based (CB) and Non-Cluster (NC) service models under 6 different scenarios and with different flow speeds and vehicle densities. CB achieves higher and more stable service ratios than NC under all scenarios and at both 6Mbps and 12Mbps data rates. The service ratio of NC also shows a raising trend with the increase of the flow speed. This is due to the lower vehicle density when the flow speed is higher, which reduces transmission collision and congestion. When the flow speed is low, the distance between vehicles is relatively short and more vehicles are involved within the same transmission range, leading to more service requests and local data collected for transmission. In this scenario, by grouping vehicles into clusters, transmission loads between vehicles and RSUs are reduced, hence less collision events in the CB model than in the NC model. When vehicles move out of the transmission range of RSU, those without support of clusters will not be able to receive service packets directly from RSU. But in the cluster-based model, CMs can still obtain services from the CH that has stored service data from RSU as long as they are in the transmission range with the CH via V2V.

The average service delay is shown in Fig. 6, which includes the time spent on transmitting service data and the waiting time for re-transmission when the previous service delivery is failed. In the NC model, each vehicle has to wait for downloading service data from the RSU in turn. This delay is reduced in the CB model since only CH is involved in V2I transmissions. In addition, more time can be saved by using a cluster where CH transmits aggregated sensing data collected from CMs and broadcasts service data from RSU to the CMs that request the same information. The delay profile presented in Fig. 6 is also correlated with the service ratio results shown in Fig. 5. When the flow speed increases, there will be less collision or congestion cases as a fewer number of vehicles are involved in transmission, thus in this scenario the CB model does not show as much advantages as they have at low flow speeds.

In Fig. 7, it is shown that the CB model clearly outperforms the NC model in terms of the network throughput under all six different scenarios. Throughput in the CB model appears to be more sustainable than that in the NC model, and the gaps between them are data rate dependant. As we can see, the CB's throughput at 6 Mbps is up to 2.3 times higher than that of the NC model, while when at 12 Mbps the difference is increased to up to 5 times. However, the throughput of the NC model also increases with the flow rate as less service requests are generated at high flow speeds or low vehicle densities.

The average throughput of individual vehicles is shown in Fig. 8 versus the flow speed. Generally, the throughput of individual vehicles in all schemes increases with the flow speed. As higher flow speeds correspond to lower vehicle densities according to Fig. 3, lower congestion in data traffic and, as a result, higher throughput will be expected in this

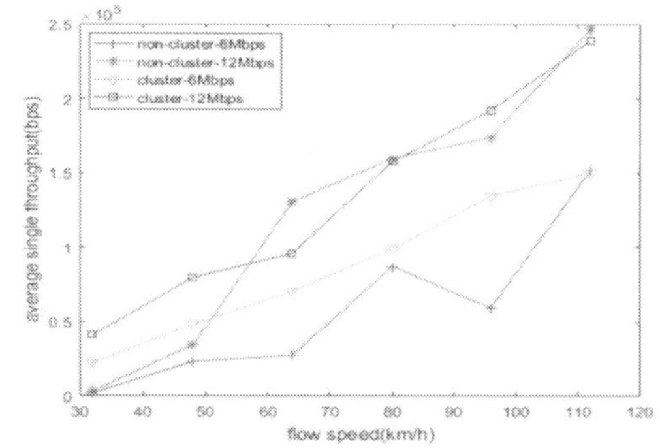

Figure 8. Average throughput of individual vehicles under different flow speeds

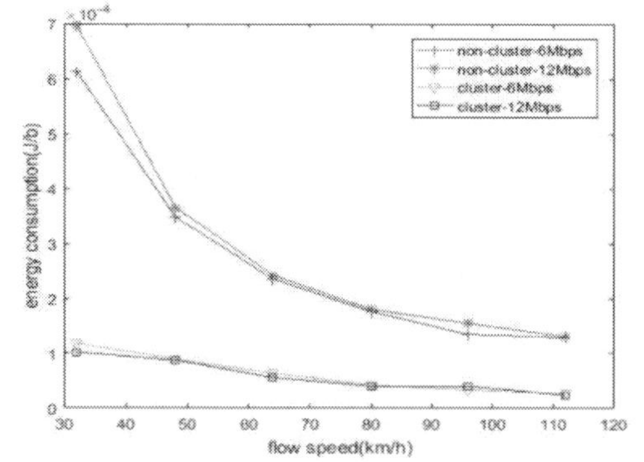

Figure 9. Energy consumption under different flow speeds

situation. In addition, it also correlates proportionally with the data rate as well. At low flow speeds, the CB model has a clear throughput advantage over the NC model because clustering helps to improve transmission efficiency. But the NC model can achieve competitively high throughput when the flow speed increases and with a higher data rate.

The energy consumption performance in terms of Joule per bit is demonstrated in Fig. 9 for the two service models. Vehicles in a cluster exchange data with a RSU via V2X, i.e. V2V between themselves and V2I between CH and RSU, while when clusters are not used all transmissions rely on V2I. This will make a significant difference in energy consumption between the two service models, as shown in Fig. 9. Like the results in other performance figures, the CB model is considerably more energy efficient than the NC model, and this advantage is particularly evident in the low flow-speed regions. The performance gap is closing down as the flow speed increases.

V. CONCLUSION AND FUTURE WORK

In this paper, we propose a service delivery model via V2X in a vehicular network to improve the transmission efficiency and reduce energy consumption. This model can effectively provide vehicles with real-time traffic and environmental information for selecting the best routes to their destinations and avoiding traffic accidents or congestions. A combined weighting metric is introduced in this paper and applied to form clusters. The CH is selected based on the mobility and connectivity of vehicles to ensure the stability and efficiency of data exchange and service delivery. As only CHs are responsible for direct communication with RSUs and dissemination of service data to other vehicles in the network, the cluster-based V2X approach presented in this work can significantly enhance service delivery efficiency and reduce energy consumption. This has been shown by simulation results, in terms of service ratio, average service delay, throughput and energy efficiency, in comparison with the performance of the V2I dominated non-cluster model.

Future work will consider more complicated scenarios in highway settings. The data aggregation method will be extended to develop specific data fusion and integration algorithms based on the information entropy theory. In addition, the two-way service model and associated energy analysis schemes will be established and investigated for developing a more realistic and efficient V2X service delivery platform.

REFERENCES

[1] A. M. Vegni, M. Biagi, and R. Cusani, Smart Vehicles, Technologies and Main Applications in Vehicular Ad hoc Networks, 2013.

[2] H. Noori and M. Valkama, "Impact of VANET-based V2X communication using IEEE 802.11 p on reducing vehicles traveling time in realistic large scale urban area," Proc. International Conference on Connected Vehicles and Expo (ICCVE), 2013, pp. 654-661.

[3] K. Liu, et al., "Network-coding-assisted data dissemination via cooperative vehicle-to-vehicle/-infrastructure communications," IEEE Trans. on Intelligent Transportation Systems, vol. 17, pp. 1509-1520, 2016.

[4] Q. Wang, P. Fan, and K. B. Letaief, "On the joint V2I and V2V scheduling for cooperative VANETs with network coding," IEEE Trans. on Vehicular Technology, vol. 61, pp. 62-73, 2012.

[5] Y. L. Morgan, "Notes on DSRC & WAVE standards suite: Its architecture, design, and characteristics," IEEE Communications Surveys & Tutorials, vol. 12, pp. 504-518, 2010.

[6] M. Jiang, J. Li, and Y. Tay, "Cluster Based Routing Protocol (CBRP). Draft-ietf-manet-cbrp-spec-01," txt, Internet Draft, IETF1999.

[7] M. M. C. Morales, C. S. Hong, and Y.-C. Bang, "An adaptable mobility-aware clustering algorithm in vehicular networks," Proc. Network Operations and Management Symposium (APNOMS), 2011 13th Asia-Pacific, 2011, pp. 1-6.

[8] I. Tal and G.-M. Muntean, "User-oriented cluster-based solution for multimedia content delivery over VANETs," Proc. IEEE International Symposium on Broadband Multimedia Systems and Broadcasting (BMSB), 2012, pp. 1-5.

[9] M. Chatterjee, S. K. Das, and D. Turgut, "WCA: A weighted clustering algorithm for mobile ad hoc networks," Cluster Computing, vol. 5, pp. 193-204, 2002.

[10] R. D. Kuhne, "Greenshields' legacy: Highway traffic," Transportation Research E-Circular, 2011.

[11] B. Association, Code of Federal Regulations, Title 47: Parts 80-End (Telecommunications): Federal Communications Commission Revised 10/05: Claitors Pub Div, 2006.

[12] D. Krajzewicz, J. Erdmann, M. Behrisch, and L. Bieker, "Recent development and applications of SUMO-Simulation of Urban MObility," International Journal On Advances in Systems and Measurements, vol. 5, pp. 128-138, 2012.

[13] C. Sommer, R. German, and F. Dressler, "Bidirectionally coupled network and road traffic simulation for improved IVC analysis," IEEE Trans. on Mobile Computing, vol. 10, pp. 3-15, 2011.

[14] The Highway Code, https://www.gov.uk/guidance/the-highway-code.

Geographic Centroid Routing for Vehicular Networks

Justin P. Rohrer
Naval Postgraduate School, Monterey, CA 93943, USA
email: jprohrer@nps.edu

Abstract—**A number of geolocation-based Delay Tolerant Networking (DTN) routing protocols have been shown to perform well in selected simulation and mobility scenarios. However, the suitability of these mechanisms for vehicular networks utilizing widely-available inexpensive Global Positioning System (GPS) hardware has not been evaluated. We propose a novel geolocation-based routing primitive (Centroid Routing) that is resilient to the measurement errors commonly present in low-cost GPS devices. Using this notion of Centroids, we construct two novel routing protocols and evaluate their performance with respect to positional errors as well as traditional DTN routing metrics. We show that they outperform existing approaches by a significant margin.**

Keywords–Vehicular network; DTN routing; Centroid routing; GPS error.

I. INTRODUCTION

Research published by the Delay- and Disruption-Tolerant Networking (DTN) community over the last decade shows significant benefits to incorporating geolocation information into routing algorithms. This is unsurprising, given that DTN routing protocols are required to make local forwarding decisions, without the benefit of consistent global routing information.

Much of this work is evaluated only in simulation and emulation environments (and we include our own prior work in making this generalization [1]), in which the positional measurements are assumed to be highly accurate. In practice, vehicular communication modules are often (and perhaps increasingly so) constructed from very inexpensive hardware without high-quality antennas or complex GPS chipsets, and expected to function in urban canyons or other environments with partially obstructed GPS signals.

Under such conditions, the advertised ± 20 m civilian GPS accuracy bounds quickly decay to nearly 200 m, with the more eccentric error typically occurring orthogonally to the direction of travel, without resembling a normal error distribution [2]. The implication for the consumer of such position data is that the location delta between updates due to error may be an order of magnitude larger than the actual distance travelled in the same time. Simply taking additional samples cannot resolve this error due to the high correlation between consecutive GPS location readings. This explains the all-too-common scenario of "my GPS thinks I'm driving in a field/lake/building/offramp/etc". Commercial GPS-based mapping devices are relatively successful at hiding such inaccuracies by taking hints from the map database and making sophisticated assumptions (learned through decades of development on this single application), such as smoothed travel trajectories and snapping the position to nearby roads. However, when these assumptions are wrong, even greater errors may be introduced so we must find other mechanisms for mitigating the underlying errors in positional data. Please note that we don't mean to imply that advances in technology won't decrease these errors; technology trickle-down, availability of Global Navigation Satellite System (GLONASS), and planned improvements in future GPS satellites will all have that effect in the coming decades, however the current general assumption of zero error will continue to be unwarranted for the foreseeable future.

Our contributions in this work include a novel routing primitive and two novel routing protocols based on this primitive. We also perform an analysis of the effects of errors in positional data on our two protocols and an existing protocol. Lastly, we contribute an oracle router for the ONE simulator. Code for all routers is made available via the Tactical Networked Communication Architecture Design lab website [3].

The structure of this paper is as follows: Section II discusses the prior work in DTN routing protocols that we build upon in this work. Section III presents our new routing primitive, and two DTN routing protocols based on that primitive. Section IV evaluates the protocols via the ONE Simulator. Section V concludes.

II. RELATED WORK

Our routing primitive design takes inspiration from a number of location-based routing protocols developed for mobile ad-hoc networks, including Vector, APRAM, DREAM, SIFT, and GRID [4]–[9]. Vector maximizes message spreading by preferentially transferring messages to neighbors traveling in a direction orthogonal to the node's own direction of travel. It calculates the trajectory vectors from repeated GPS samples. APRAM [10] utilizes GPS coordinates to discover the geographically shortest path to the destination, while DREAM uses the cached node locations to make local forwarding decisions that forward packets in the direction of the destination. Similarly, AeroRP [11][12][13] uses both the coordinates and velocity of neighbors to locally determine the best next hop. LAR [14] uses location information to bound the area of the route discovery phase, thus reducing overhead. Beaconless geographic routing [15] exploits the broadcast nature of wireless channels to overhead the location of neighboring nodes, and use this information to discover the best route. Other protocols such as IGF [16], BOSS [17], and BLR [18] have been proposed that vary in the algorithm used to select the forwarding node.

We presented a 2-page poster paper describing the issue of ignoring GPS errors when simulating geographic grouting protocols [19], however it did not include our novel protocols or the simulation analysis described in this work.

III. Centroid-Based Routing

In this work, we introduce a novel geographic routing primitive called the *Centroid*. In physics, the centroid is defined as "the center of mass of a geometric object of uniform density" and our *Centroid* intentionally evokes this idea in the context of geographic routing. In looking at the location history of a mobile node, whether a circuit, linear path, or other arbitrary trace, we can envision a central point at which that trail would be balanced. Unlike the physical centroid, which takes into account all the mass composing an object, we only concern ourselves with the points on the trace itself, which indeed may not for a closed shape at all. This centroid then may be calculated as:

$$C_x(t_p) = \sum_{t=1}^{t_p} \frac{C_x(t-1) \times (t-1)}{t} + \frac{x_t}{t} \tag{1}$$

Where C is the Centroid, x is the X, Y, or Z component of the node position, and t_p is the present time increment. Time increments are in terms of the chosen update interval. In practice, we calculate the delta between the old Centroid and the new Centroid at each update as follows:

$$\Delta C_x(t_p) = \frac{x_{t_p} - C_x(t_p - 1)}{t_p} \tag{2}$$

This primitive is naturally resistant to noise introduced into the location history due to GPS reception errors, since they will be averaged out over time. This is in contrast to positional routing primitives that rely only few/recent GPS readings to make routing decisions, and in some cases apply transformations such as trajectory calculation that amplify the effects of errors in those readings.

As with probabilistic and other routing protocols that predict a node's behavior based on past locations and encounters, the primary assumption with the Centroid is that a node's behavior will have repetitive qualities, so it is not suitable for one-shot type mobility patterns. To go from primitive to protocol, then, there are many possible paths. We explore a couple of these is this paper and leave others to future work.

A. Goals

We have a few goals for our protocols:

1) Resilience to noise/error found in positional data
2) Low power consumption
3) Simplicity; at this point we want to evaluate the Centroid primitive, not overshadow it with complex behaviors

The first and third are relatively self-explanatory, but the second deserves some additional discussion. While there are many factors that influence power-consumption, in this context we assume that avoiding unnecessary packet transmissions will have the greatest impact. We also observe that some DTN protocols are explicitly designed to maximize use of resources in order to improve the probability of delivery messages. This is entirely reasonable given certain assumptions, such as vehicular networking where power is essentially unlimited with respect to communications. Under other assumptions, such as personal devices and low-power sensors this approach is not optimal. In our case, we are designing for the later case, which also corresponds to devices that commonly have cheap GPS receivers and compromised antennas. Another consideration is that op-in users are less likely to forward packets if it noticeably drains the battery on their device. For these reasons we want to explicitly conserve resources where possible.

B. Centroid Router

One mechanism that has shown significant promise in prior investigations is that employed by the Vector routing protocol, where the number of messages exchanged at an encounter is proportional to the orthogonality of the two nodes' trajectories. Unfortunately, as described earlier, projecting a trajectory amplifies the effects of positional measurement errors. We propose the Centroid-based analog of this, where the number of messages exchanged at an encounter is proportional to the cartesian distance between the two nodes' Centroids. We call the implementation of this the Centroid Router.

At each encounter, the Centroid Router exchanges its current centroid, message list (list of all messages currently in the node's buffer), ACK list (list of all messages that have been acknowledged by the recipient), and current neighbor list. ACK'd messages are deleted from each nodes' buffer first, and message exchange is begun with delivery of messages addressed to the nodes exchanging the messages. The next set of messages exchanged are those addressed to currently connected neighbors of the two nodes. None of these messages count against the message limit. If this is all completed and the link is still established, the nodes proceed to exchange other messages subject to the message limit. To find the limit, each node calculates the Centroid Distance, which is the distance between its own Centroid and the Centroid of the node it is exchanging messages with. It then finds the ratio between this distance and the longest Centroid Distance it has previously calculated for all node encounters. The resulting ratio is the fraction of its own message list that it is allowed to send to the connected neighbor. This results in a linear relationship between Centroid Distance and fraction of messages exchanged. More messages are exchanged with nodes that have more distant centroids, thus spreading messages as widely as possible with fewer forwarding events.

C. CenterMass Router

The CenterMass Router begins with the Centroid Router, and adds one additional mechanism, which is to forward messages *in the direction of* their destination. To achieve this, each node maintains a list of the Centroids for all nodes it has encountered, along with the address of each node and timestamp of the encounter. At each encounter, Centroid lists are exchanged and merged into the nodes' own Centroid list. In the case of a collision for a particular node address, the node keeps only the newer entry based on the timestamps.

When forwarding messages, the message limit is calculated as in the Centroid Router, however messages are only forwarded if the distance between the neighbor's Centroid and the destination's Centroid is smaller than the distance between the current host's Centroid and the destination's Centroid. This mechanism minimizes the spread of messages in a direction away from the destination.

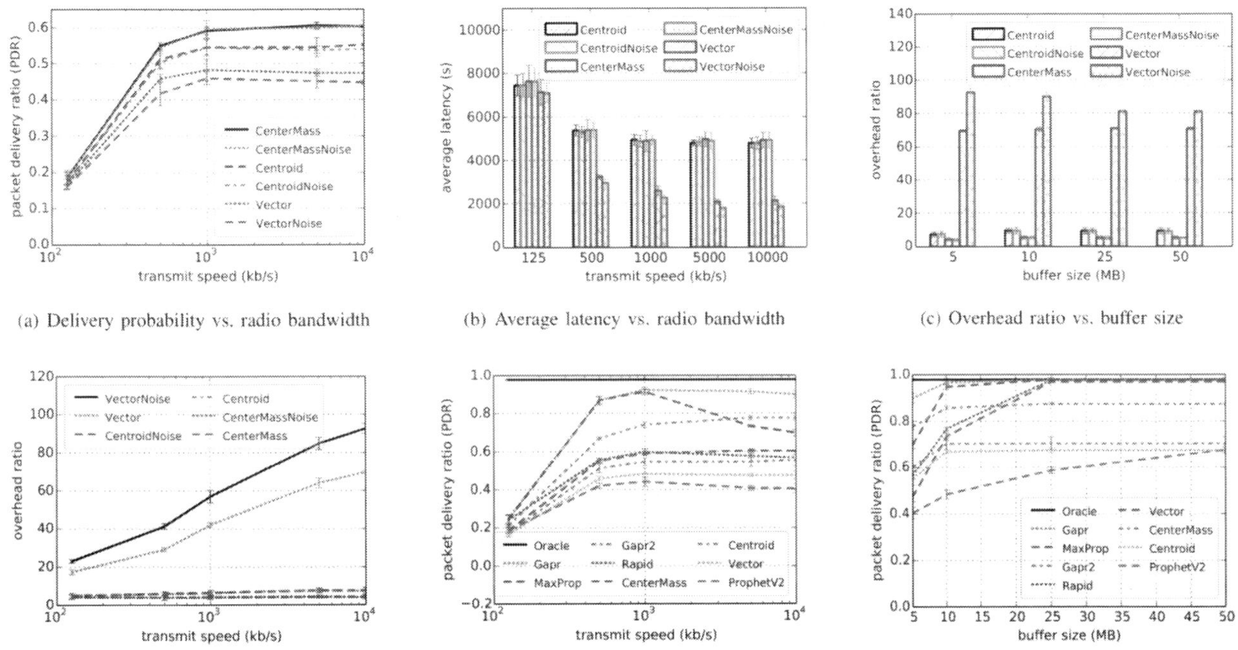

(a) Delivery probability vs. radio bandwidth

(b) Average latency vs. radio bandwidth

(c) Overhead ratio vs. buffer size

(d) Overhead ratio vs. radio bandwidth

(e) Delivery probability vs. radio bandwidth

(f) Delivery probability vs. buffer size

Figure 1. Performance effects of GPS errors

TABLE I. HELSINKI SCENARIO PARAMETERS

Parameter	Value
simulated duration	12 hrs
warmup time	1000 s
timestep resolution	0.1 s
number of runs	4
radio bandwidth	10 Mb/s
transmit range	10 m
buffer size	5 MB
number of pedestrians	80
pedestrian speed	0.5–1.5 m/s
pedestrian pause time	0–120 s
number of cars	40
car speed	2.7–13.9 m/s
car pause time	0–120 s
number of trams	6
tram speed	7–10 m/s
tram pause time	10–30 s
message rate	1 / 25–35 s
message size	0.5–1.0 MB
message TTL	5 hrs

IV. SIMULATIONS & ANALYSIS

We perform our analysis using The ONE Simulator [20], as it is specifically suited to DTN routing analysis. In previous work, we have used a number of mobility scenarios with the ONE, however in this case evaluating the protocols on multiple scenarios did not yield any additional insights, so for clarity we present a single evaluation scenario in this work. We choose the Helsinki map-based model, which has become well-known in DTN routing literature due to its inclusion as the default mobility model for the ONE simulator. The model includes both vehicles and pedestrians that participate as nodes in the routing protocol. We have made some minor changes to the default parameter values, which are shown in Table I. To generate traffic, one random node sends a message to a random destination every 25–35 seconds. Our simulation study consists of three components:

1) Evaluating the effect of positional measurement errors
2) Comparing our protocols' performance to more sophisticated probabilistic protocols that rely on encounter history to predict path costs
3) Discussing design tradeoffs and attempting to quantify their effect

Each data point in the plots that follow represents the average of 4 simulation runs with varying random seeds, and the error bars on all the plots in this paper represent 95% confidence intervals.

A. Positional Sample Error

As discussed earlier, we are concerned with the effect of errors in the positional (e.g. GPS) sample data provided to the routing protocol. To evaluate this, we create alternate versions of both of our protocols (Centroid and CenterMass) as well as the Vector routing protocol, which add noise to the position provided by the simulator before using it in calculating their respective routing primitives. This noise is random and uniform, in the range ± 20 m. Note that this roughly the *advertised* error for civilian GPS, and is far from a worst-case scenario that could be ± 200 m with strong correlation between samples.

Figure 1(a) shows how this measurement error affects the Packet Delivery Ratio (PDR) of the three protocols. All the

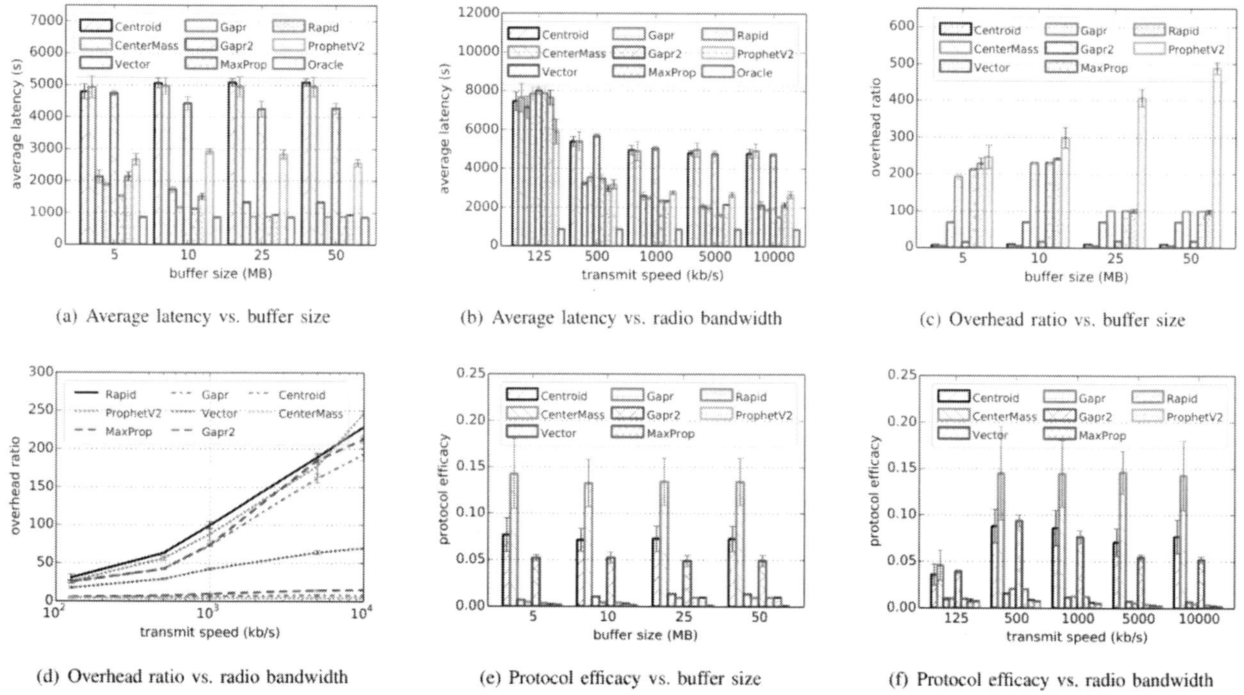

(a) Average latency vs. buffer size

(b) Average latency vs. radio bandwidth

(c) Overhead ratio vs. buffer size

(d) Overhead ratio vs. radio bandwidth

(e) Protocol efficacy vs. buffer size

(f) Protocol efficacy vs. radio bandwidth

Figure 2. Helsinki scenario performance metrics

protocols are bunched together at low (125 Kb/s) radio data rates. With higher transmission rates (500 Kb/s – 10 Mb/s) the protocols become distinguishable. The Vector protocol is most significantly affected, with the positional errors noticeably reducing the packet delivery ratio. That being said, the reduction is only about 10% at its worst. We believe that since the Vector protocol only relies on the trajectory to enable efficient spreading of messages, it may be less affected than a protocol that uses trajectory in a more specific manner (e.g. identifying trajectory in the direction of the message destination). Unfortunately, we do not have such a protocol implemented in the ONE at this time to test our hypothesis. Both of the Centroid-based protocols show negligible effects from the noise, as expected. While the traces with noise trend lower than those without, they are within the 95% confidence intervals of each other at almost every data point. Not only is the Centroid routing protocol less affected by noise, it outperforms the Vector protocol by about 20% in the presence of positional errors. The CenterMass protocol achieves an additional 10% performance improvement over Centroid.

We next examine the effect on latency, shown in Figure 1(b). The effects are very small across the board, but surprisingly Vector's latency *improves* in the presence of errors. Vector also outperforms the Centroid-based protocols and higher transmission rates. Lastly, we look at the effect on overhead. The overhead ratio reported the the ONE simulator is: $\frac{(\text{forwarded messages}-\text{delivered messages})}{\text{delivered messages}}$. From Figure 1(c) we can now explain the reduced latency achieved by the Vector protocol, since there are literally 10× more copies of every packet forwarded in the Vector routing simulations that there are in the Centroid routing simulations, and the positional errors make the Vector overhead approximately 30% worse. Increasing the buffer size reduces the impact of positional errors on Vector's overhead, and has almost no effect on Centroid or CenterMass. We do note that in addition to the improved delivery probability of CenterMass over Centroid, CenterMass has significantly lower overhead than Centroid. For a view of the effects of increasing transmission speed we look to Figure 1(d). Here, we see that not only does the effect of positional errors on Vector increase as more bandwidth is made available, but the absolute overhead appears to run-away, quadrupling between 125 Kb/s and 10 Mb/s. Centroid also has increased overhead as the bandwidth increases, but only slightly, and there appears to be almost no effect on the overhead of CenterMass.

From these plots, we see that the negative impact of positional errors on some protocols is real, and that these two Centroid-based routing protocols have a significant advantage in terms of overhead, relative to Vector, a protocol of comparable complexity and message delivery performance.

B. Probabilistic-Predictive Routing Comparisons

Having satisfied the question of whether the Centroid primitive minimizes the effects of positional error, we can then proceed to comparisons with a larger selection of routing protocols in under ideal (no GPS error) circumstances confident that our protocols will not degrade with real positional data and offering a number of other protocols a best-case scenario of no errors (and indeed several of them do not rely on positional data anyway). Here, we are looking to see how our relatively simple protocols compare to much more sophisticated protocols that use encounter history to predict probability of future delivery, including a couple from our own prior work (Geolocation Assisted Routing Protocol (GAPR) and GAPR2) [21]. For this comparison we have also created an Oracle router that demonstrates the best possible delivery ratio given the scenario contact graph. Figure 1(e) shows these results, with respect to radio transmission rate. One feature of particular note, is that while most protocols reach a maximum delivery probability and then plateau, MaxProp [22], though it ties for the highest overall delivery ratio, decreases significantly (20%) with *higher* transmission rates. We do not have an explanation for this, but have observed it in the past. Aside from this anomaly, MaxProp and our own GAPR and GAPR2 protocols all out perform CenterMass and Centroid, by as much as 30%. CenterMass is roughly equivalent to Rapid [23], both of which are about 10% improved over Centroid, with Vector and PRoPHETV2 10 and 20% worse than Centroid, respectively. Taken by itself, this shows that the sophistication of probabilistic protocols is not without merit. Figure 1(f) paints a similar picture, but also shows that with increase available buffer space some of the lesser performing protocols in the previous plot show great improvement, notably Rapid and Vector, both of which approach the performance of the Oracle with sufficient buffer availability. Unfortunately, neither of the Centroid-based protocols fall into this category, and their only redeeming quality in this plot is that they continue to outperform PRoPHETv2 [24]. Fortunately, that is not the end of the analysis!

Pressing on we examine latency in Figures 2(a) and 2(b). Here, we see that CenterMass and Centroid are at the higher-end of the latency spectrum, along with GAPR2, while our older protocol GAPR can acheive approximately the same latency as the Oracle, given sufficient buffer space. We note that MaxProp, Vector and Rapid all perform well with respect to latency. We also note again the trend of improvement with increased buffer sizes, which leads us to the topic of overhead.

Examining Figure 2(c) we see that the Centroid-based protocols have overhead ratios of 10 or below, and are invariant to buffer size. With the exception of GAPR2, the other protocols are about 1 order of magnitude higher, with PRoPHETv2's overhead increasing dramatically at higher buffer sizes. In Figure 2(d) we see that Vector's runaway overhead was relatively minor compared to most of the other protocols whose overhead appears to increase exponentially in some cases as transmission rate increases. As before, the overhead of Centroid and CenterMass are small and relatively unaffected by the radio bandwidth.

C. Protocol Efficacy

Given that the message delivery performance of our Centroid-based protocols was outclassed by several of the probabilistic protocols, but they were the clear leader in terms of overhead, we would like to be able to show the combined effect using a single metric that captures the goals we presented earlier on in the paper. We would use efficiency, but it is roughly the inverse of overhead, and does not fully serve our purposes. So, we introduce Protocol Efficacy, which is simply $\frac{(\text{delivery ratio})}{\text{overhead}}$. In the ideal case where delivery ratio is 1 and overhead is 1, efficacy would also be 1, and any time delivery ratio is 0, efficacy is 0. This metric then captures both the goal of high delivery ratio and low overhead. Figures 2(e) and 2(f) show this metric for each of the protocols discussed so far. We see that the CenterMass protocol outperforms the rest of the field by a significant margin, Centroid and GAPR2 are comparable to one another, and the rest (Vector, GAPR, MaxProp, Rapid, and PRoPHETv2) fall behind by and order of magnitude or more.

D. Observations

The simulation of the Rapid routing protocol runs about two orders-of-magnitude slower than any of the other protocols tested. Since the total number of messages forwarded in the Rapid simulations is not particularly high, we can only assume that the computations required are of significant complexity. This may be of concern when trying to conserver power on low cost/performance devices.

We again note that some routing protocols are designed to continue transmitting as long as connection stays up, even in the face of diminishing marginal performance gains, based on the assumption that power (and therefore transmissions) are free. This design choice results in the runaway overhead results seen above, but may not be of concern for certain environments. Our design is different in that it explicitly stops transmitting to conserve resources when those transmissions are unlikely to result in message delivery. This then reflects the difference is design philosophies between explicitly conserving resources (power) and maximizing resource use (bandwidth/buffer space).

V. Conclusion

We have demonstrated the negative effect that positional errors can have on routing protocols in a vehicular network that rely on geolocation inputs, depending on how that input is used. We have also demonstrated Centroid Routing and CenterMass Routing, both of which are immune to random error in positional data inputs. We have shown how these protocols out-perform existing state-of-the-art probabilistic routing protocols both in terms of traditional metrics and using our new Protocol Efficacy metric. These new protocols show a dramatic improvement over existing protocols when normalized against the overhead they induce in the network.

We envision many possible applications of the Centroid primitive. One of the simplest is as a direct substitute for position in position-based routing. Another is as an enhancement to our own GAPR2 routing protocol, to replace the raw position data currently employed. The CenterMass routing protocol also has potential for refinement, and we think that by combining geolocation and encounter history data in some meaningful way we can approach the delivery probability of probabilistic protocols while retaining the Efficacy of the Centroid-based protocols shown here. We expect that the Efficacy metric will be particularly significant in evaluating high-volume, loss tolerant routing environments such as drone-swarms [25]. In the future we intend to continue this work using a simulator with higher-fidelity network, MAC, and physical layer models, such as ns-3, in which we have begun implementing DTN routing protocols [26].

Acknowledgment

This work was funded in part by the US Marine Corps and US Navy. Views and conclusions are those of the authors and should not be interpreted as representing the official policies or position of the U.S. government.

References

[1] J. P. Rohrer, A. Jabbar, E. Perrins, and J. P. G. Sterbenz, "Cross-layer architectural framework for highly-mobile multihop airborne telemetry networks," in Proceedings of the IEEE Military Communications Conference (MILCOM), San Diego, CA, USA, November 2008, pp. 1–9.

[2] B. Ben-Moshe, E. Elkin, H. Levi, and A. Weissman, "Improving accuracy of GNSS devices in urban canyons," in Proceedings of the 23rd Canadian Conference on Computational Geometry (CCCG), August 10–12 2011, pp. 399–404.

[3] Center for tactical networked communications architecture design. https://tancad.net. [retrieved: July, 2018]

[4] H. Kang and D. Kim, "Vector routing for delay tolerant networks," in Proceedings of the IEEE 68th Vehicular Technology Conference, Sept 2008, pp. 1–5.

[5] W.-H. Liao, J.-P. Sheu, and Y.-C. Tseng, "GRID: A fully location-aware routing protocol for mobile ad hoc networks," Telecommunication Systems, vol. 18, no. 1-3, 2001, pp. 37–60.

[6] M. G. de la Fuente and H. Ladiod, "A performance comparison of position-based routing approaches for mobile ad hoc networks," Vehicular Technology Conference (VTC), October 2007, pp. 1–5.

[7] L. Galluccio, A. Leonardi, G. Morabito, and S. Palazzo, "A MAC/routing cross-layer approach to geographic forwarding in wireless sensor networks," Ad Hoc Netw., vol. 5, no. 6, 2007, pp. 872–884.

[8] M. Mauve, A. Widmer, and H. Hartenstein, "A survey on position-based routing in mobile ad hoc networks," IEEE Network, vol. 15, no. 6, 2001, pp. 30–39.

[9] M. Yuksel, R. Pradhan, and S. Kalyanaraman, "An implementation framework for trajectory-based routing in ad hoc networks," Ad Hoc Networks, vol. 4, no. 1, 2006, pp. 125–137.

[10] M. Iordanakis et al., "Ad-hoc routing protocol for aeronautical mobile ad-hoc networks," in Proceedings of the Fifth International Symposium on Communication Systems, Networks and Digital Signal Processing (CSNDSP), 2006, pp. 1–5.

[11] A. Jabbar and J. P. G. Sterbenz, "AeroRP: A geolocation assisted aeronautical routing protocol for highly dynamic telemetry environments," in Proceedings of the International Telemetering Conference (ITC), Las Vegas, NV, October 2009, pp. 1–10.

[12] J. P. Rohrer, A. Jabbar, E. K. Çetinkaya, E. Perrins, and J. P. Sterbenz, "Highly-dynamic cross-layered aeronautical network architecture," IEEE Transactions on Aerospace and Electronic Systems (TAES), vol. 47, no. 4, October 2011, pp. 2742–2765.

[13] J. P. G. Sterbenz et al., "Disruption-tolerant airborne networks and protocols," in UAV Networks and Communications, 1st ed., K. Namuduri, S. Chaumette, J. H. Kim, and J. P. G. Sterbenz, Eds. Cambridge University Press, January 2018, ch. 4, pp. 58–95.

[14] Y.-B. Ko and N. H. Vaidya, "Location-aided routing (LAR) in mobile ad hoc networks," Journal of Wireless Networks, vol. 6, no. 4, Jul. 2000, pp. 307–321.

[15] J. Sanchez, P. Ruiz, and R. Marin-Perez, "Beacon-less geographic routing made practical: Challenges, design guidelines, and protocols," IEEE Communications Magazine, vol. 47, no. 8, August 2009, pp. 85–91.

[16] B. Blum, T. He, S. Son, and J. Stankovic, "IGF: A state-free robust communication protocol for wireless sensor networks," Department of Computer Science, University of Virginia, USA, Technical Report CS-2003-11, 2003.

[17] J. Sanchez, R. Marin-Perez, and P. Ruiz, "BOSS: Beacon-less on demand strategy for geographic routing in wireless sensor networks," in IEEE Internatonal Conference on Mobile Adhoc and Sensor Systems (MASS), October 2007, pp. 1–10.

[18] M. Heissenbüttel, T. Braun, T. Bernoulli, and M. Wälchli, "BLR: Beacon-less routing algorithm for mobile ad hoc networks," Computer Communications, vol. 27, no. 11, 2004, pp. 1076–1086.

[19] J. P. Rohrer, "Effects of GPS error on geographic routing," in Proceedings of the 26th International Conference on Computer Communications and Networks (ICCCN). Vancouver, Canada: IEEE, August 2017, pp. 1–2.

[20] A. Keränen, J. Ott, and T. Kärkkäinen, "The ONE simulator for DTN protocol evaluation," in Proceedings of the 2nd International Conference on Simulation Tools and Techniques (SIMUTools). New York, NY, USA: ICST, 2009, pp. 55:1–55:10.

[21] J. P. Rohrer and K. M. Killeen, "Geolocation assisted routing protocols for vehicular networks," in Proceedings of the 5th IEEE International Conference on Connected Vehicles (ICCVE), Seattle, WA, September 2016, pp. 1–6.

[22] J. Burgess, B. Gallagher, D. Jensen, and B. N. Levine, "Maxprop: Routing for vehicle-based disruption-tolerant networks," in Proceedings of the 25th IEEE International Conference on Computer Communications (INFOCOM), April 2006, pp. 1–11.

[23] A. Balasubramanian, B. Levine, and A. Venkataramani, "DTN routing as a resource allocation problem," in Proceedings of the ACM conference on Applications, technologies, architectures, and protocols for computer communications (SIGCOMM), vol. 37, Oct. 2007, pp. 373–384.

[24] S. Grasic, E. Davies, A. Lindgren, and A. Doria, "The evolution of a DTN routing protocol – PRoPHETv2," in Proceedings of the 6th ACM Workshop on Challenged Networks (CHANTS). ACM, Sep. 2011, pp. 27–30.

[25] A. Pospischil and J. P. Rohrer, "Multihop routing of telemetry data in drone swarms," in Proceedings of the International Telemetering Conference (ITC), Las Vegas, NV, October 2017, pp. 1–10.

[26] J. P. Rohrer and A. N. Mauldin, "Implementation of epidemic routing with ip convergence layer in ns-3," in Proceedings of the 2018 Workshop on ns-3 (WNS3). Surathkal, India: ACM, June 2018.

PhyCoNet-Sim: A Framework for Physically Accurate Simulations of Vehicular Ad-Hoc Networks

Steffen Moser, Ralf Schleicher and Frank Slomka
Institute of Embedded Systems/Real-Time Systems
Faculty of Engineering, Computer Science and Psychology
Ulm University, Albert-Einstein-Allee 11, 89081 Ulm, Germany
Email: steffen.moser@uni-ulm.de, ralf.schleicher@uni-ulm.de, frank.slomka@uni-ulm.de

Abstract—**For decades, the simulation has been a well-established methodology to study the behavior of wireless telecommunication networks. While network and link layer protocols are simulated by using very detailed models, there is typically still a lack of explicit and deterministic considerations of the physical layer. Phyical layer, antenna and radio channel are regularly approximated with simplistic models based on statistic distribution of bit error rate values. While this approach might deliver a sufficient accuracy to compare the performance of routing protocols or other higher-level applications, it disallows, however, studying cross-layer concepts, multi-user communication or the explicit consideration of deterministic channel models. It also denies the physical layer to be an explicit degree of freedom in the design space. To overcome this problem, we set up a PhyCoNet-Sim, a co-simulation framework which links OMNeT++ to a physical layer simulator based on GNU Radio or Matlab/Simulink.**

Keywords–VANET; Simulation; Physical Layer; Matlab; Simulink

I. INTRODUCTION

Besides field tests and formal verification approaches, simulations are an important methodology to understand the behavior of distributed systems. Especially in the research area of wireless communications in highly dynamic scenarios like Vehicular Ad-hoc NETworks (VANET), simulations are a very effective method to get an in-depth understanding of a complex network's behavior: Compared to field tests, simulations offer fast and reproducible results in an early design phase without the necessity to develop, setup and deploy the systems. In highly mobile scenarios, also formal verification approaches reach their limits: A mobile network's behavior relies heavily on stochastic processes, e.g., mobility behavior, radio channel conditions, and so on. To allow formal verification methodologies to be applied to these systems, a lot of data would have to be gathered and statistically evaluated in the first place by exploiting measurement campaigns. For this reason, simulations play a well-established and major role in VANET research.

The wireless network simulation frameworks, which are typically used today, come with very accurate and deterministic models of the network and link layer protocols. The reason is very obvious: The protocol behavior can be simulated efficiently by using discrete event-triggered simulation engines. Highly performance-optimized frameworks like OMNeT++ [1], NS-3 [2] or JiST [3] rely on this technique. Specialized vehicular ad-hoc network extensions are available, e.g., Vehicles in Network Simulation (VEINS) [4] or Scalable Wireless Ad-Hoc Network Simulator (SWANS). They allow simulating even large-scale scenarios within a reasonable amount of computation time. The simulated link layer or network protocol code can be similar or even identical to implementations used in real-world systems, so aspects like handshaking, queuing and forwarding packets, sending acknowledgments, finding optimal paths, updating routing tables and evaluating properties like channel load, network capacity and packet delivery rates can be done very realistically.

When looking into the internals of state-of-the-art simulation frameworks, one can clearly identify a major problem: The system's components which reside below the link layer (e.g., the components of the physical layer (PHY), the antenna(s) and the radio channel) are modeled in a very simplistic way: A scalar Bit Error Rate (BER) value is used as the only environmental input data to the protocol simulation when the simulator has to decide about the reception of a frame. Using a look-up table, the BER is derived from the Signal-to-Noise Ratio (SNR) value, which is itself based on a fixed transmission power, the distance of the communicating nodes, a stochastic fading model and a background noise level, for example Additive White Gaussian Noise (AWGN). Abstractly spoken, the BER value used by the link layer simulation represents the whole behavior of physical layer, antennas and radio channel with having only the communicating nodes' spatial distance as an environmental input. Slightly improved simulation frameworks extend this approach by using the Signal-to-Interference-plus-Noise Ratio (SINR) to estimate the BER. In SINR-based models, the signal powers of neighboring network nodes are added to the background noise level.

The reasons for using the above-mentioned simplistic models may be found among the following challenges: Physically accurate channel models and signal-based physical layer models do not fit well into discrete event-triggered simulations: The physical layer introduces signal representations, which have to be considered at least in a time-triggered manner, the physics of antenna and radio channel clearly belong to a continuous-time domain. Considering these models explicitly consumes much more computation time compared to classical simulations. Another difficulty is caused by the fact that PHY implementations are done in hardware, so the network protocol designers do not have access neither to the used algorithms nor to their implementations. The simplifications, which are applied, often reach through the whole simulation frameworks: Taking an arbitrary IEEE 802.11p simulation in OMNeT++ as an example, a situation is implicitly considered as a collision at the potentially receiving nodes if more than one node gets into transmission mode in the same time interval within a certain spatial range. While this behavior is the correct one for protocols using a Carrier-Sense Multiple Access with Collision Avoidance (CSMA/CA) strategy on the Medium Access Control (MAC) layer, it renders many other approaches impossible [5].

The remainder of this paper is structured as follows: In Section II, the limitations of today's ad-hoc network simulators are illustrated. In Section III, we introduce literature where possibilities are discussed to overcome these limitations. Section IV is used to present our approach of co-simulating radio channel, PHY layer and network. In Section V, we evaluate our approach. We close this paper with Section VI, where we summarize and give a perspective to future work. In Section VII, we offer a download link to the framework.

II. Problem Statement

The following list gives an excerpt of four aspects which are hard to explore with state-of-the-art network simulators:

- The state-of-the-art Dedicated Short-Range Communication (DSRC) protocol in vehicular ad-hoc networks is based on IEEE 802.11p. Technically, IEEE 802.11p is an adaption of IEEE 802.11a to vehicular networks, while the IEEE 802.11a standard has been published in 1999. Both, Wireless Local Area Networks (WLAN) and cellular networks have evolved a lot since then, the main reason is that higher-integrated circuits allow much more advanced signal processing schemes. This means, Multiple-Input, Multiple-Output (MIMO) transceivers, Multi-User MIMO (MU-MIMO) transceivers and directed beam forming are common in today's IEEE 802.11ac, which is ubiquitous in homes and offices. The upcoming IEEE 802.11ax standard will contain even more interesting advances in the PHY layer, so there is a distinct interest to evaluate the suitability of these advances for VANETs, especially since limitations of IEEE 802.11p are becoming more and more evident [6]. This is a difficult task with classic network simulations. It would require to gather a lot of statistics data in advance, but even this makes it hard to simulate the behavior realistically because the channel's influence (multi-path propagation, shadowing by buildings) is much higher than in 802.11a or 802.11p. Therefore, it is more difficult to accomplish everything with a single statistics-based distribution. When considering beam forming, there is a cross-layer scenario between PHY layer and network layer [7], [8].
- It is an open question where to place antennas on vehicles. Especially MIMO systems, which necessarily introduce multiple antennas per node, require the number, the form and the position of antennas to be explicit properties of the design space. To determine, for example, which configuration is optimal in specific environmental scenarios (e.g., free-space intersection, high buildings, partly shadowed antennas by trucks), it is necessary that the electrical behavior of antennas to be modeled.
- In ad-hoc network simulations, the SNR or SINR values are calculated by a model of the radio channel physics. The radio channel models used in state-of-the-art simulations are typically rather simple combinations of stochastic fading models on the one hand and distance-based path-loss models on the other hand. The latter parts determine the SNR by an implementation of Friis transmission equation [9] which has been published in 1946. Besides the Euclidean distance of the two communication nodes, Friis' formula considers the gains of the antennas and their effective size based on the wavelength. It does not consider environmental effects like buildings, vehicles and plants. There are, of course, also more sophisticated models available. It has been shown that in highly dynamic network topologies, like VANETs, simple distance-based path-loss models do not offer the required accuracy. This is especially true for urban scenarios where buildings introduce unequally distributed shadowing effects. This has been extensively studied in [10], [11], and [12]. For this reason, ray-optical channel models have been developed which calculate a delay spread of a signal based on the three-dimensional environment. For example, in vehicular ad-hoc networks buildings, vehicles and the terrain roughness are considered by ray-optical approaches. The resulting delay spread shows the temporal diversification and power distribution of different multi-path components. To link ray-optical channel models to traditional wireless network simulators, the delay spread needs to be reduced to a scalar SNR value, effectively discarding most of its information.
- Cellular networks of the fifth generation (5G) are ready to be deployed. The cellular network service providers promise remarkably lower latencies and higher data rates compared to 4G. Due to the fact that cellular network-based applications in vehicles' comfort and emergency systems already exist, the influence of cellular network-based services will presumable increase further. To evaluate a combined usage of cellular and DSRC communication or study protocol convergence concepts in VANET scenarios, a simple IEEE 802.11p-tailored simulation framework is not sufficient anymore.

We present the framework PhyCoNet-Sim, which uses co-simulation of radio channel, physical layer and the upper layers of the distributed system in order to tackle down the restrictions, which lead to the above-described problems. All components are already existing as open-source or commercial tools. For this reason, we don't present a more advanced simulation or approach for a given subcomponent in this paper. Instead, we combine the best approaches available to a co-simulation by designing and implementing suitable interfaces.

III. Related Work

In this section, we present work which is related to the approach presented in this paper. Note that our approach to make the PHY layer an explicit variable in the design phase of a distributed system is intentionally suitable for wireless communication systems in general. However, VANETs are among these kinds of applications, which make the most challenging demands due to the low channel coherence times caused by the high mobility of the transmitting and receiving but also of the neighboring nodes. For this reason, we compare our approach especially to related work in VANET research.

The necessity of enhancing both precision and capabilities of wireless ad-hoc network simulations is not new. Our challenge papers date back to 2010 [13]. The ideas presented were driven by problems which occurred when trying to integrate results of deterministic channel models into VANET simulations. Deterministic channel models typically generate an impulse response, i.e., a distribution of signal power over time which shows the delay-spread of an Dirac impulse and thus the channel's properties. VANET simulators did not offer any interface for an impulse response, as the statistics-based models work with the scalar BER value, only.

In the meantime, different concepts have been proposed. In [14], Papanastasiou et al. present a method for integrating the simulation of the physical layer into network simulations. Their approach is based on the discrete event simulator NS-3 and uses the IT++ library for the transceiver implementation. This approach does not fully conform to the IEEE 802.11p standard since it solely implements the OFDM transmission method, the PHY frame format, the modulation and the coding schemes. As channel model either the path loss or the Rayleigh fading model were employed, and only a rudimentary traffic simulation method was used which did only encode relative movement and did not take shadowing and other environmental effects into account. This simulator is named PhySim-Wifi for NS-3, is available in the version 1.2, and accessible on the website of the research group. Apparently, there are no more releases since April 2012 and we, therefore, consider the project to be discontinued.

In [15], Judd and Steenkiste describe a Hardware-In-the-Loop-based (HIL) channel emulator. Using a wired connection between the antenna ports of real wireless hardware and a channel emulator based on a Field-Programmable Gate Array (FPGA), they can emulate both, the channels effects on the radio propagation and real PHY behavior. The system supports movement of the simulated nodes by altering the channel conditions accordingly. Interfering signals are superposed. They use a simple path-loss model as a channel model, but the authors point out that simulation of the 3D environment is possible by implementing, for example, a ray-tracing based model on the FPGA. This is a very interesting and flexible approach for accurate physical layer consideration in wireless network simulations: Real network interface hardware is used for signal processing. At the same time this is a major downsides of this concept: It obviously depends on hardware implementations of the protocol, which is about to be studied. Wireless network interfaces must be already available which renders the approach unsuitable in a very early design phase.

Making statistics-based BER calculation more accurate is also still a topic in communication technology research. In [16], Schneider et al. did a large measurement campaign, but used fixed base stations as known from cellular networks. A more detailed campaign, which addresses especially the channel behavior in VANET, was done by Walter in 2016 [17]. He showed that the VANET scenario has a strong impact on choosing the correct statistics distribution.

In [18], Bloessl et al. describe their design and implementation of a Software-Defined Radio (SDR) of IEEE 802.11p using the open-source SDR framework GNU Radio. It has been verified by comparison against commercially available IEEE 802.11p hardware. It contains a complete PHY layer especially for DSRC applications.

Based on Bloessl's work, we developed *Signal Simulation* in 2015, which represents an interface between VEINS and OMNeT++ on the one hand and GNU Radio on the other hand [19]. It was shown that firstly, it works successfully and secondly, for IEEE 802.11p the simplifications present in VANET simulators are valid – at least, when neglecting channel effects and explicit PHY simulation. It is of special interest that the approach is generic and not necessarily bound to IEEE 802.11p, which is only used for proof of concept. This approach solved our requirements from a theoretical point of view completely. Unfortunately, the GNU Radio framework lacks models for the most-recent highly-sophisticated WLAN standards like IEEE 802.11n/ac/ax or for the most recent cellular network standards. When looking at PHY models besides the IEEE WLAN standards, for example, 4G/LTE, one finds models which have been partly developed.

To sum up, using an SDR for co-simulation is the way to go. Having a pure open-source GNU Radio implementation is a promising application to gain deterministic models for the generation and decoding of signals, but has a major drawback regarding the available building blocks.

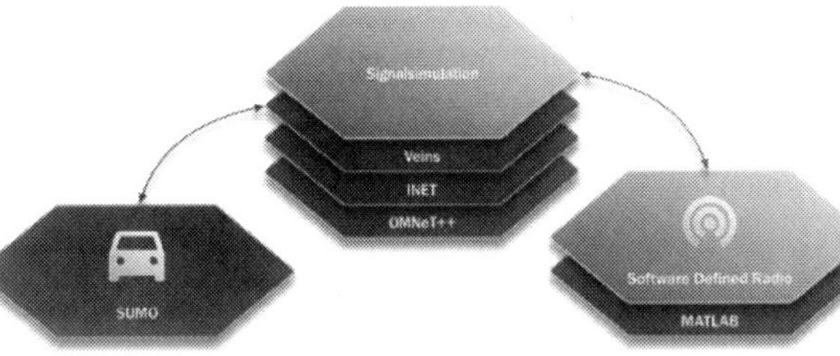

Figure 1. Schema of tools used in PhyCoNet-Sim

IV. PhyCoNet-Sim: Channel, Physical Layer and Network Co-Simulation using a Software-Defined Radio Approach

In this section, we present our approach PhyCoNet-Sim – a co-simulation of channel, physical layer and the network. Essentially, it is an extension of Signal Simulation to allow a flexible substitution of the subcomponents. Figure 1 gives an overview of the components used by PhyCoNet-Sim: VEINS and OMNeT++ are well-accepted ad-hoc network simulators. Simulation of Urban Mobility (SUMO) [20] is a widely-used microscopic vehicular traffic simulator. Matlab/Simulink [21] by The Mathworks is an extremely popular platform in different kinds of engineering disciplines. It comes with a lot of toolboxes. The Communication System Toolbox [22], the WLAN System Toolbox [23] and the Antenna Toolbox [24] exactly address all aspects of the physical layer and the antenna. Systems generated by the help of these toolboxes are specified in a very detailed way, because with the help of additionally available code generators, the toolboxes are actually used in designing and deploying SDRs. For our purpose, we can directly exploit the SDR code to simulate the physical layer. In contrast to GNU Radio, the above-mentioned Matlab/Simulink toolboxes come with implementations of a lot of highly-configurable IEEE WLAN standards, even components of upcoming standards are supported. Regarding radio channel modeling, the Communication System Toolbox comes with a lot of statistics-based methods. As there is a real signal representation for all transmitted packets in the simulation, there is a common interface for integration of impulse responses, which can stem from measurements, statistics-based models and ray-tracing-based models. For the latter, we implemented parts of [10], [12], [25] and [26] to allow deterministic channel behavior for a given environmental input.

In the next sections, the most important components of our PhyCoNet-Sim approach are described. We especially stress the differences between Signal Simulation and PhyCoNet-Sim.

A. *Transmission*

The MAC layer and all layers above the MAC layer are simulated as in a classical VEINS simulation. We assume that a packet is generated by one of the upper layers, e.g., a Cooperative Awareness Message (CAM) is to be sent. In this case, an according event is generated and queued. The simulation engine jumps to the point in time where the event will take place. As we use CSMA/CA in our IEEE 802.11p reference implementation, it will be checked whether the ready-to-transmit node detects a free channel. If so, the transmission process takes place. Note that the approach works also for non-CSMA MAC protocols, in case of Code-Division Multiple Access (CDMA) or MU-MIMO approaches, other possibilities to detect when a transmission can be scheduled are possible. Whenever a transmission event takes place, for all receivers within a configurable maximum distance, a reception event is triggered. In a classical VEINS simulation, the receivers within a certain range would get a copy of the transmitted data depending on the BER based on the SNR value. The latter would be depending on the distance of the nodes to estimate the long-term path loss and a stochastic process to estimate the short-term fading. This does not happen in PhyCoNet-Sim.

B. Reception

1) Signal Generation: PhyCoNet-Sim does not deliver the data to the receiving nodes. Instead, in a first step, the position vectors and velocity vectors of the transmitting and the receiving vehicles are stored. This information is necessary as input data for the channel model. A detailed schema of the reception process is depicted in Figure 2. After storing speed and position of the vehicles, the signal, which corresponds to the transmitted frame, is generated. Therefore, the transmitter part of the Matlab-based SDR modem is called via the interface developed for this work. It maps the transmitter's data frame to a signal constellation which is a discrete-time set of baseband signal samples. In an SDR setup these signal samples would be fed to a digital-to-analog converter and later to an up-sampler, a transmission amplifier and to the antenna. In our case, we store the baseband time-domain signal samples in a matrix.

2) Antenna Characteristics and Radio Channel Influence: In order to simulate the system as accurately as possible, the simulated signal must suffer from the antenna's and the radio channel's influence. This is put in execution by convolving the generated transmitter signal with the channel's impulse response. The latter can be generated by a deterministic ray-optical channel model, for example. Using the impulse response defines a very versatile interface which allows different kinds of channel and antenna models to be applied. Instead of using the ray-optical model, it is also possible to use statistics-based channel models available in Matlab's Communications System Toolbox as well as data sets gathered by measurement campaigns [16], [17]. As the velocity and location profiles of the transmitting node and the receiving node are known, it is possible to map the channel effects caused by the movement directly to a non-stationary channel response. This means: Our model allows to consider a node's movement during the transmission process which needs, of course, a non-zero amount of time to transmit a frame of a certain length. The granularity, which is available therefore, is defined by the mobility model used in the simulation. In our test-case, we used vehicular traffic models available in SUMO because there is already a very good integration in VEINS. All signal-processing parts are done in Matlab/Simulink by using the introduced toolboxes. For this reason, the user has access to a large number of building blocks and implemented wireless communication standards. It is also possible to model and simulate complex multi-antenna scenarios using the Antenna Toolbox and consider their behavior.

3) Signal Buffering: After convolving the generated signal with the channel's impulse response, the resulting signal pattern contains all multi-path, delay, Doppler-spread, antenna-gain and AWGN noise information. In a two-node scenario or when using a Time-Division Multiple Access (TDMA) MAC protocol, the signal pattern could be directly forwarded to the Matlab SDR's receiver routine. In a multi-hop scenario with all nodes being unsynchronized this is not possible. Although this process can be very memory-consuming, it is necessary to store the resulting signal pattern in a signal buffer. The reason is that we have to consider other neighboring nodes, which might be transmitting in a time interval which overlaps with the time interval of the first signal we received. Even with a collision avoiding strategy, this can happen due to the high mobility of the nodes. If we simulated only CSMA-based protocols, we could consider such effects as collisions and discard the received signal. In our scenario, we are going to feed the contents of the signal buffer to a Matlab SDR's receiver engine, which allows us to do an unbiased decoding of the signal. Before triggering the receiver engine, other incoming signals have to be treated iteratively as explained here. All overlapping parts are stored in the signal buffer. Compared to our GNU Radio-based implementation, the data structure used for the signal buffer in this work is adaptive. Instead of providing a fixed amount of memory, which can be costly when it comes to large scenarios, the memory is now allocated dynamically.

4) Interferences: Interference is a local phenomenon of wave physics. This means that the superposition of interfering waves has to be calculated for each receiving node. To do so, we use the signal components, which are stored in a node's signal buffer. In our SDR approach, we do not have a mathematical function of a continuous-time signal in our buffer which we could simply add-together. While the channel and antenna effects on the signal could be simulated in a fully continuous-time domain (e.g., by using a computer algebra system), the signal, which leaves our SDR transmitter, consists of discrete-time samples which is the way to go for real systems deployments. For this reason, also the signal in our buffer at the receiver's site is described by samples instead of a closed mathematical function. To avoid any loss of generality, an accurate PHY layer simulation requires that a receiver model is able to cope with signal components which can arrive at arbitrary points in time. Signals, which reach a receiver, can originate from different senders being active at overlapping time intervals or from multi-path propagation. Caused by the finite signal propagation speed of the signals, both cases introduce an arbitrary delay, which is independent from the simulator event steps or any time steps. The SDR models in Matlab do not solve the calculation of the superposition directly. For this reason, interpolation approaches have been implemented as we already proposed in [19]. While linear time interpolation would be very easy to apply to all signal samples in the signal buffer, it introduces an error. In our tests with the Matlab SDR, the rate of reception with respect to a given signal-to-noise ratio was worse when linear interpolation was used compared to using the nearest sample, in which case there was no interpolation. This confirms our previous results. A better solution is an ideal band-limited interpolation by using a fractional-delay filter [27].

5) Decoding: The superposition calculated by the fractional-delay filter is forwarded to the receiver part of the Matlab SDR. It tries to decode the superposition of the different signal components, channel-introduced distortions and the background noise. Depending on the channel coding parameters, a specific amount and distribution of bit errors may be corrected in the PHY layer. After a successful decode, the node's MAC layer will be notified about an incoming data frame. For more advanced decoders (e.g., MU-MIMO in the up-link as proposed for IEEE 802.11ax), there can be several decoding processes, which can take place at the same time.

C. Software-Defined Radio

The SDR modem used for our implementation has been configured by using the WLAN System Toolbox of Matlab

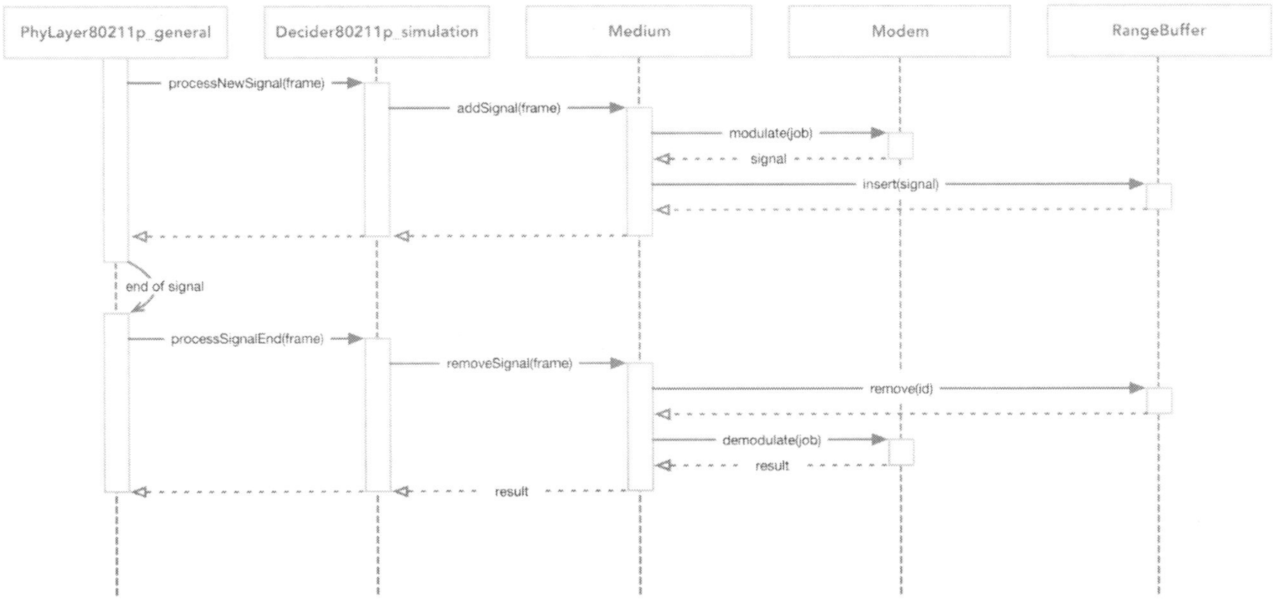

Figure 2. Detailed schema of a signal reception process in PhyCoNet-Sim

2017a. Essentially, the WLAN System Toolbox generates and configures Matlab code for transmitter and receiver. The following supported modem types are supported:

- Type:
  - IEEE 802.11a/b/g/p (Non-High Throughput)
  - IEEE 802.11n (High Throughput)
  - IEEE 802.11ac/ax (Very-High Throughput)
- Channel bandwidths: 10, 20, 40, 80 and 160 MHz
- Modulation schemes: Direct-Sequence Spread Spectrum (DSSS) and Orthogonal Frequency Division Mutliplexing (OFDM)
- Coding schemes:
  - BPSK 1/2, BPSK 3/4, ...
  - QPSK 1/2, QPSK 3/4, ...
  - QAM-16 1/2, QAM-16 3/4, ...
  - QAM-64 2/3, QAM-64 3/4, ...
  - ...

The parameters can be mixed, which means different coding schemes can be combined with various channel bandwidths. The code generated by the WLAN System Toolbox is Matlab code. Thus, it is easily possible to extend it by user-code to evaluate new approaches, which go beyond the implemented standards. Other wireless systems besides WLAN can be designed by using the building blocks from Communications System Toolbox. As a comparison, GNU Radio essentially covers the non-high throughput modems. The execution of the transmitter and receiver functions within the Matlab interpreter offers interesting possibilities for probing signals, but is inefficient. For this reason, we use the Matlab Coder to generate native C++ code when we link it to a network simulation. Essentially, the Matlab coder generates two C++ functions: One for the transmitter and one for the receiver. These functions are linked via an interface to our signal buffer layer in OMNeT++.

V. EVALUATION

In this section, we describe the first evaluation steps of the implemented PhyCoNet-Sim. In a first step, we started with an implementation of our GNU Radio-based approach of [19]. We compared it with PhyCoNet-Sim regarding accuracy and performance.

A. Accuracy

To get a an impression regarding the accuracy, we took the IEEE 802.11p and compared the GNU Radio receiver with the Matlab receiver by simulating the probabilities of a successful signal frame reception for different signal-to-noise ratios. The plot is depicted in Figure 3 and shows a very similar trend. The Matlab SDR shows a higher reception quality, i.e., it reaches better decoding probabilities for SNR values between 23 dB and 35 dB. We are currently in the process of investigating the differences.

B. Performance

Highly-accurate physical layer simulations consume a lot of computation time compared to traditional network simulations. Essentially, the major part of the costs of computation depend on the number of signal transmissions and receptions. Assuming that one frame is sent which is received by n neighboring nodes, there is one transmission and n receptions. Another cost driver are the length of a frame and the amount of the delay-spread caused by the channel: A large temporal delay-spread of a signal in the signal buffer causes more discrete-time signal samples which must be considered by the decoder engine. In the GNU Radio approach, a signal consisting of 64 OFDM symbols lead to a computation of 115 ms for a whole transmit-receive-cycle (1 transmit, 1 receive). GNU Radio uses Python for connecting the flow graph of signal processing modules. In the GNU Radio approach, the Python interpreter has been restarted each time a signal is received. We optimized our Matlab SDR approach regarding performance. On a similar computer architecture like it was used for our

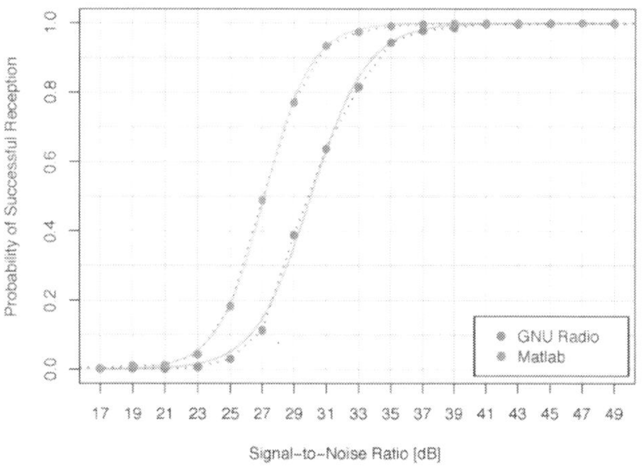

Figure 3. Signal reception probability for the GNU Radio and the Matlab IEEE 802.11p receiver

GNU Radio approach, we got at least an improvement by a factor of 11.

VI. Conclusion and Future Work

The Communication System Toolbox, the WLAN System Toolbox and the Antenna Toolbox offer modern physical layer models as parts of the Matlab/Simulink tool suite. While their primary focus is the development of software-defined radios, we have shown in this paper that the models can be applied very suitably to complex vehicular ad-hoc network simulation scenarios. Therefore, we enhanced an existing approach, which already linked OMNeT++ with GNU Radio, by the ability to delegate the PHY layer calculation to the Matlab toolboxes. We have implemented a prototype for the simulation of IEEE 802.11a/g/p and provided an integration in the VEINS simulator. However, our approach is both, protocol agnostic and application agnostic, i.e., all parts of the co-simulation can be exchanged. In future work, we will use the developed framework for performance simulations for more modern PHY layer mechanisms in vehicular ad-hoc networks. Another interesting research question will be radio convergence (5G and WLAN) in complex applications.

VII. Availability

The PhyCoNet-Sim framework has been integrated into the *WAVE – Next Generation* project. It is available at http://wave-ng.net.

References

[1] "OMNeT++," retrieved: 2018-04-30. [Online]. Available: https://omnetpp.org/

[2] "The Network Simulator NS-3," retrieved: 2018-04-30. [Online]. Available: http://www.nsnam.org/

[3] R. Barr, "An Efficient, Unifying Approach to Simulation Using Virtual Machines," Ph.D. dissertation, Cornell University, 2004.

[4] "Vehicles in Network Simulation (VEINS)," retrieved: 2018-04-30. [Online]. Available: http://veins.car2x.org/

[5] B. Bloessl, F. Klingler, F. Missbrenner, and C. Sommer, "A Systematic Study on the Impact of Noise and OFDM Interference on IEEE 802.11p," in 9th IEEE Vehicular Networking Conference (VNC 2017). Torino, Italy: IEEE, November 2017, pp. 287–290.

[6] B. Bloessl, M. Gerla, and F. Dressler, "IEEE 802.11p in Fast Fading Scenarios: From Traces to Comparative Studies of Receive Algorithms," in 22nd ACM International Conference on Mobile Computing and Networking (MobiCom 2016), 1st ACM International Workshop on Smart, Autonomous, and Connected Vehicular Systems and Services (CarSys 2016). New York, NY: ACM, October 2016, pp. 1–5.

[7] H. Stuebing, A. Jaeger, N. Wagner, and S. A. Huss, "Integrating Secure Beamforming into Car-to-X Architectures," SAE International Journal of Passenger Cars- Electronic and Electrical Systems, vol. 4., Jun. 2011, pp. 88–96.

[8] S. Moser, S. Eckert, and F. Slomka, "An Approach for the Integration of Smart Antennas in the Design and Simulation of Vehicular Ad-Hoc Networks," in Proceedings of the International Conference on Future Generation Communication Technology (FGCT), London, UK, Dec 2012, pp. 36–41.

[9] H. Friis, "A Note on a Simple Transmission Formula," in Proceedings of the I.R.E. and Waves and Electrons, vol. 41, 1946.

[10] J. Maurer, W. Sörgel, and W. Wiesbeck, "Ray Tracing for Vehicle-to-Vehicle Communication," in In Proceedings of the URSI XXVIIIth General Assembly 2005, New Delhi, India, Oct. 2005.

[11] I. Stepanov, D. Herrscher, and K. Rothermel, "On the Impact of Radio Propagation Models on MANET Simulation Results," in Proceedings of the 7th IFIP International Conference on Mobile and Wireless Communication Networks (MWCN 2005), Marrakech, Morocco, pp. 61–78.

[12] S. Moser, F. Kargl, and A. Keller, "Interactive Realistic Simulation of Wireless Networks," in Proceedings of the IEEE/EG Symposium on Interactive Ray Tracing 2007, 2007, pp. 161–166.

[13] S. Moser and F. Slomka, "Towards more Realistic Simulations of Ad-hoc Networks - Challenges and Opportunities," in Proceedings of the International Symposium on Performance Evaluation of Computer and Telecommunication Systems (SPECTS), Juli 2010, pp. 422–427.

[14] S. Papanastasiou, J. Mittag, E. G. Strom, and H. Hartenstein, "Bridging the Gap between Physical Layer Emulation and Network Simulation," in Wireless Communications and Networking Conference (WCNC), 2010 IEEE. IEEE, 2010, pp. 1–6.

[15] G. Judd and P. Steenkiste, "A Software Architecture for Physical Layer Wireless Network Emulation," in Proceedings of the 1st International Workshop on Wireless Network Testbeds, Experimental Evaluation & Characterization. ACM, 2006, pp. 2–9.

[16] C. Schneider, M. Narandžić, M. Kaske, G. Sommerkorn, and R. Thoma, "Large Scale Parameter for the WINNER II Channel Model at 2.53 GHz in Urban Macro Cell," in Vehicular Technology Conference (VTC 2010-Spring), 2010 IEEE 71st, May 2010, pp. 1–5.

[17] M. Walter, "Scattering in Non-Stationary Mobile-to-Mobile Communications Channels," Ph.D. dissertation, Ulm University, 2016.

[18] B. Bloessl, M. Segata, C. Sommer, and F. Dressler, "Towards an Open Source IEEE 802.11p Stack: A Full SDR-based Transceiver in GNURadio," in 5th IEEE Vehicular Networking Conference (VNC 2013). Boston, MA: IEEE, December 2013, pp. 143–149.

[19] D. Maier, S. Moser, and F. Slomka, "Deterministic Models of the Physical Layer Through Signal Simulation," in Proceedings of the 8th International Conference on Simulation Tools and Techniques, ser. SIMUTools '15, 2015, pp. 175–182.

[20] "SUMO - Simulation of Urban MObility," retrieved: 2018-04-30. [Online]. Available: http://sumo.dlr.de/

[21] The Mathworks, "Matlab," retrieved: 2018-04-30. [Online]. Available: https://mathworks.com/

[22] ——, "Communications System Toolbox," retrieved: 2018-04-30. [Online]. Available: https://mathworks.com/products/communications.html

[23] ——, "WLAN System Toolbox," retrieved: 2018-04-30. [Online]. Available: https://mathworks.com/products/wlan-system.html

[24] ——, "Antenna Toolbox," retrieved: 2018-04-30. [Online]. Available: https://mathworks.com/products/antenna.html

[25] A. Schmitz and L. Kobbelt, "Wave Propagation Using the Photon Path Map," in PE-WASUN '06: Proceedings of the 3rd ACM international workshop on Performance evaluation of wireless ad hoc, sensor and ubiquitous networks. ACM Press, 2006, pp. 158–161.

[26] J. Nuckelt, M. Schack, and T. Kürner, "Deterministic and Stochastic Channel Models Implemented in a Physical Layer Simulator for Car-to-X Communications," Advances in Radio Science, vol. 9, no. C. 5-1, 2011, pp. 165–171.

[27] V. Valimaki and T. I. Laakso, "Principles of fractional delay filters," in 2000 IEEE International Conference on Acoustics, Speech, and Signal Processing. Proceedings (Cat. No.00CH37100), vol. 6, 2000, pp. 3870–3873 vol.6.

Toward Safety and Security Development by Identifying Interfaces of Automotive Functions

Toru Sakon and Yukikazu Nakamoto
Graduate School of Applied Informatics
University of Hyogo
Kobe, Japan
Email: tr.sakon@gmail.com, nakamoto@ai.u-hyogo.ac.jp

Abstract—**An item is a system or an array of systems used to implement a function at the vehicular level. However, owing to increasing demands for advanced functions and security features in an automobile, an item and the manner in which it is defined has become more complex. In this research, we propose using resource sharing as the basis for defining an item and its boundary. We use four simple categories for introducing an influencer that represents a shared resource and its management function. This makes the process of defining an item simple and straightforward. Further, by refining an influencer, complex interaction between systems in an item is better described. We applied this method to define an item for the sample systems.**

Keywords–ISO 26262; item definition; interface; resource sharing; management; security.

I. INTRODUCTION

Safety in case of malfunction or failure of a vehicular system is a priority in every vehicle. To ensure safety, the functional safety approach is adopted in the development of an electrical and/or electronic (E/E) system in a vehicle. ISO 26262 is a widely adopted international standard for ensuring functional safety of an E/E system in a vehicle [1]. In ISO 26262, the development targets are termed as items. Items are defined by their functions at the vehicular level. Definition of an item boundary is essential for an item. However, in the current advanced vehicular functions, there are some cases in which an item boundary cannot be clearly defined. The functions at the vehicular level are realized using more than one in-vehicle Electronic Control Unit (ECU). However, in the case of an ECU shared by multiple items, interference between the items may result via shared resources of the ECU. For advanced vehicular level functions, multiple items are closely integrated to realize the function. The development of cybersecurity measures is indispensable to the development of safety functions, considering the cyberattacks on the safety functions. In this context, during the development of security functions on the basis of items, the characteristics of attack points are necessary for defining items. However, security attack points for the items are not necessarily included in the items defined by in-vehicle function and their boundaries. In other words, for the development of secure advanced functions in vehicles, we need a sophisticated method to formalize all interactions among the target items. Reflecting this fact, compositional aspects are added to the item definition in the latest draft standard [2]. However, an interaction based model of a combined system requires tightly-coupled processes calling each other's interfaces. This increases the complexity in designing combined items. In this study, we introduce a new object called an influencer to maintain conventional item definitions. An influencer defines shared resources and management functions for the items. With the introduction of an influencer, we propose a method to incorporate the newly introduced high-functional impacts and security considerations as an interface requirement into the item definition while maintaining the granularity of a conventional item definition.

The structure of this paper is organized as follows. Section II briefly describes the difficulties in designing combined items for automotive safety and security. We state the basic idea of an influencer and use cases of the designing procedure with an influencer, in Section III and IV, respectively. Section V describes related works. We present conclusion in Section VI.

II. DEVELOPMENT OF FUNCTIONAL SAFETY IN AN AUTOMOTIVE E/E SYSTEM

The functional safety design of an E/E system is central to safety design of an automobile. Functional safety means "the absence of risks due to hazards caused by the malfunctioning behavior of E/E systems" [3]. ISO 26262 is a functional safety standard for automotive electronic systems. Its development process is defined based on a V-shaped process. Item definition starts in the early stage of the development process (See Fig. 1).

An item is defined as "a system or array of systems to implement a function at the vehicular level, to which ISO 26262 is applied" [3]. The boundary of an item with that of other items is an item's important property. It is defined considering a) elements of the item; b) environment of the item; c) interactions of the item with other items or elements; d) functionality required by other items; e) functionality required

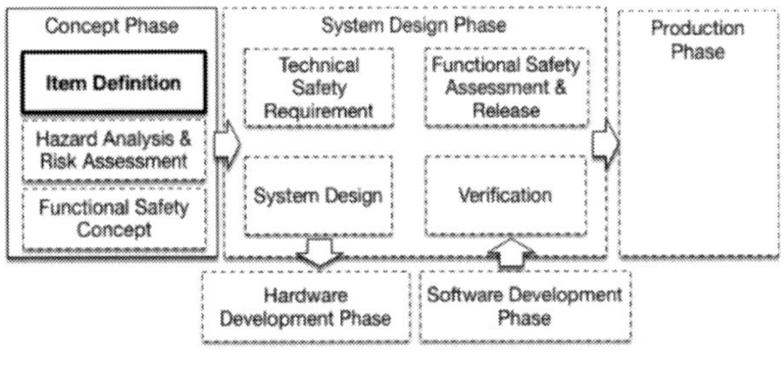

Figure 1. Development process of ISO 26262 and the target of this study (bold line)

from other items; f) allocation and distribution of functions; and g) operating scenarios [1].

However, in general, defining a highly complex system by different subsystems and determining the boundary and interaction between them is a difficult task. In addition to the complexity of the system itself, the complexity of mutual communication between the elements of the complex system is very high. Therefore, the structure of interactions between the subsystems is highly dependent on the way, in which a complex system is subdivided. Further, it is difficult to eliminate all the hidden communication paths in a complex system. An automobile is a highly complex system and hence, it is difficult to define an item and its boundaries during its development. We categorize these difficulties into three types. The first category is the definition and description of boundaries related to potential sharing, particularly non-message-like associations that do not appear explicitly. For example, consider an item consisting of one or more systems. Furthermore, another item shares one or more of its systems. In this case, there is an interaction between two items via shared systems. A typical example is sharing an ECU with multiple items. In this case, there are some mutual effects owing to sharing of resources, such as memory, Central Processing Unit (CPU) usage, Input / Output (IO), or basic software resources. However, sharing of ECU resources is not expressed explicitly. Therefore, it is difficult to identify items to incorporate item definition.

The second problem is defining sophisticated item boundaries during the differential development of complicated functions. A simple example is given in [4] but for advanced vehicle functions, multiple items cooperate to realize the function. For example, a Lane Keeping Assist System (LKAS) and a Parking Assist System (PAS) are realized by integrating several vehicular level functions, such as a steering, throttle control, braking, and the functions that control them. From the aspect of reusing a proven system, it is preferable for a developer of differential development to minimize and isolate the side-effects of the new function from the preexisting product. Furthermore, for the development based on item definitions, items should be isolated by their boundaries as clearly as possible. To satisfy both these objectives, one of the rational ways is to handle predefined items as shared resources, define the management of item, and finally introduce the newly defined shared resource and its management function.

The third problem is developing an item definition method considering the factor of cybersecurity. One of the serious threats to car cybersecurity is the scenario where the vehicle safety function is compromised by cyberattacks. A malicious attacker must search the attack path to the function first, in order to compromise the vehicle safety function. If the attacker finds the path, he can begin the attack on the function. In ISO 26262, the safety functions of vehicles are studied on an item and its boundary basis. The attack path to the safety functions should be through the item boundary. Therefore, it is reasonable to subject an item and its boundary to an analysis of car cybersecurity. This means that, in threat analysis, influencers are candidates for attack surfaces for items. The entry point of a threat is assumed to be located at an influencer. Threat analysis method will be applied to the item and related influencer for the assessment of the threat. For cyberattack against items, there are direct attacks on functions defined in items and indirect attacks, such as attack on resources used by functions. In indirect attacks, as per the current item definition in ISO 26262, the management is possibly not considered as an item because resource management is not directly related to vehicular level functions. As a result, "feature" is introduced as a subject of security instead of an item in [5]. Furthermore, there are cyberattacks from the attack point that they are not included in the item. For example, in an in-vehicle network, a cyberattack may be possible from a compromised ECU, which is not included in the item. Therefore, there is a need for a method to explicitly describe cybersecurity requirements in the item of functional safety development.

III. Solution to the Problems

The problem categories exist due to the difficulties of considering the indirect communication between items. The first and third category of problems arise from not considering indirect communication via ECU hardware, basic software, or architecture. The second category of problems arises from difficulties in describing the complex communication patterns of advanced functions between the items which are based on mutual communication between systems belonging to an existing product. In other words, these problems are due to the interface being examined from the aspect of direct interaction. An example of relationship that cannot be found only from the decomposition of functions is of two independent systems sharing a Controller Area Network (CAN) network. These two systems have no interfaces to each other. However, once one of the systems is compromised and starts a Denial of Service Flooding attack on the network, the other system is also affected. In order to realize a function which consists of multiple elements, we propose a method to define the item and its boundary based on the interaction between the shared resources and their management. By explicitly describing this shared structure, we aim to organize and add potential relationships between items as functional additions and resources that can be added to items and reflect them in the development after item definition. This newly introduced structure is called an influencer. First, resource sharing among items is classified by resources while their management method by four simple categories, i.e. categories of influencers. The types of resources are logical resources (information) and physical resources (hardware, physical resources etc.). Resource management includes transferring and sharing of resources (TABLE I). Each cell of TABLE I is a definition of an influencer between items. TABLE I lists four categories of influencers.

- Movement of information resources corresponds to data movement between items. The information may consist of a command, message, or other data. An example of a management mechanism is a communication protocol.
- Sharing information resources is equivalent to referring the same data from multiple items. Examples of management functions of shared memory are exclusive control, distributed shared memory management protocol, and operating system resource control.
- Movement of physical resources is accompanied by movement of physical objects, for example, electric

power transfer during charging. An example of management mechanism is power delivery control.

- Sharing of physical resources implies sharing media and resources, such as memory, communication media, buses, and power supply. The management methods include bus arbitration and time division control.

In this paper, the boundaries of items are defined as follows:

1) Define shared resources. If prerequisite architecture is available, it may be used to define shared resources.
2) Define the way of management of shared resource to fit into one of the categories listed in TABLE I. This is the initial description of the influencer.
3) Refine the initial description of the influencer to assign it to each item. In this procedure, the influencer may be divided into sub-influencers and define the interaction between them.
4) Allocate the defined (sub) influencer to each item. In other words, the influencer is defined as an element that describes sharing and management of information, data, physical quantities, and their mechanisms among the items.

IV. USE CASES

In this section, we provide some use cases.

1) In case of the definition and description of boundaries related to potential sharing, we must identify their hidden interaction. We suppose the implementation of two items on a single ECU as an example (Fig. 2). Simply sharing an ECU i.e. hardware, implies sharing its hardware resources, such as CPU, available memory, and IO systems (Fig. 2 a). Next, we define the management of these resources. In most ECUs, the CPU is managed using a time division approach and memory is managed by a preassigned fix region and a dynamically allocated region. The shared IO between two items is managed in an exclusive manner (Fig. 2 b). Thus, we have three candidates for the influencer. We now focus on the IO system. The IO resource is managed exclusively; for an exclusive control, we need a protocol between resource requester (in this case, an item) and provider (in this case, the influencer). At this point, we can refine the influencer as the combination of two influencers. One is the management of the IO itself and the other is accessing data and its management protocol (Fig. 2 c).

2) In case of the item definition and its boundaries in differential development, we need to define them to minimize the modification from predefined items. In this study, we assume a PAS consisting of a steering system, brake system,

TABLE I. CATEGORIES OF INFLUENCERS

	Transferred	Shared
Logical resources (Information)	Communication (Message transfer, Remote Procedure Call)	Data Sharing (Shard memory, Shared object) or Code (Function)
Physical resources	Physical Transfer (Battery Charging)	Physical Sharing (Communication bus, Battery)

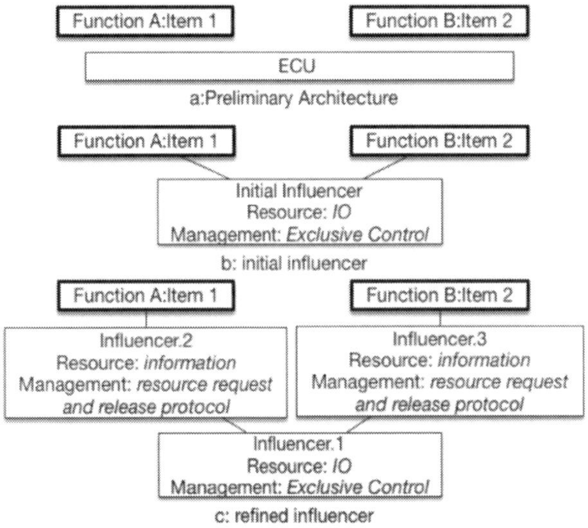

Figure 2. An ECU shared by items

throttle system, and a PAS controller system (Fig. 3). We assume that the first three systems are already well defined as preexisting products and they consist of a user interface system and a control system for the control target. In other words, each item of a preexisting product consists of user interface and controller functions (Fig. 3 a). In this case, the user interface or controller functions share control of the steering, brake, and throttle system. Then, we define status data of a PAS shared among the items and management of its exclusive control. These are the resources and controls required for the initial description of the influencer (Fig. 3 b). Next, for refinement of the influencer, the initial influencer is decomposed into two parts namely, the shared status and consistency control of it. These parts are assigned to each item of the predefined product and the newly defined PAS controller system (Fig. 3 c).

3) In case of the item definition method considering cybersecurity, we need to define an item and its boundary to identify potential cyberattacks to it. We consider some items connected by a network as an example (Fig. 4). In this case, we assume a CAN network (Fig. 4 a). The network itself is physically shared. Further, the data frames on the network are logically shared and managed following the priority and arbitration rule. In other words, this is the initial definition of the influencer (Fig. 4 b). A malicious attacker can attack a target item by compromising another item connected to the network. To take this attack into consideration, we decompose the initial influencer into two categories. One is the network influencer, whose resources are the network and bus arbitration. The other is the node influencer that sends and receives information from the network influencer. In this scheme, an item and its boundary have the node influencer as their boundary. A malicious attack can be formulated using the malicious data received from the network. Thus, by adding the node influencer for item definition, we can take cyberattack from other item as receiving malicious data on node influencer (Fig. 4 c). Moreover, changing the management of a network influencer, the attack condition on a node influencer may be changed. For example, dividing a network into two sub-networks and

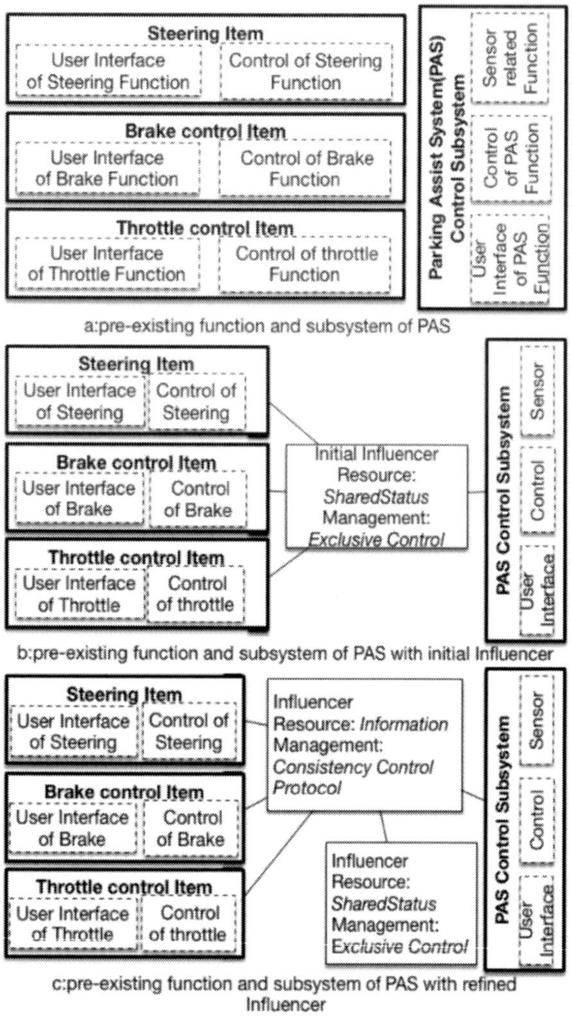

Figure 3. Advanced complex item

connecting them by a new item such as a gateway ECU, attacks from the other sub-network may be restricted (Fig. 4 d).

V. Related Works

There are considerable previous studies on the functional decomposition of the system as well as studies that perform hierarchical decomposition on the functional basis for vehicles [6]. However, it is necessary to assume anomalies and attacks from parts that do not directly have a functional relationship from the system complexity and response to cyberattacks in the future. Such a relationship is not included in the relationship between functions. In this research, we infer that the method of functional decomposition focused on the sharing side is more comprehensive than that of based on the relation between the functions.

VI. Conclusion and Remarks

In this study, we proposed a new item definition method by introducing an influencer. Particularly, focusing on logical and physical resources shared by item boundaries and their management, we proposed a method to extract the construction elements of influencers using four simple categories. We formulate that by sharing resources, remote cyberattacks are also captured by the granularity size of conventional items. In the definition of sophisticated functions that consolidate multiple items, the proposed method facilitates decomposing items. In the future, we would like to examine the effectiveness and designing technique of the multi-layered defense of an in-vehicle system with an influencer as the key.

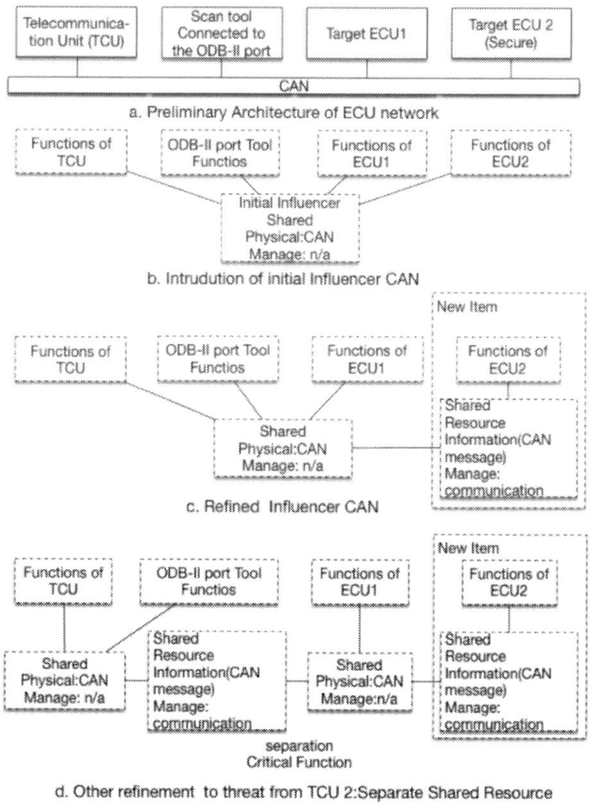

Figure 4. Item definition of a secure in-vehicle system

Acknowledgment

This work is partly supported by JSPS KAKENHI Grant Numbers 16H02800 and 17K00105.

References

[1] ISO, Road vehicles - Functional safety - Part 3: Concept phase , ISO 26262-3:2011 (E).

[2] ISO, Road vehicles - Functional safety - Part 3: Concept phase , ISO/FDIS 26262-3:2018 (E).

[3] ISO, Road vehicles - Functional safety - Part 1: Vocabulary, ISO 26262-1:2011 (E).

[4] ISO, Road vehicles - Functional safety - Part 10: Guideline on ISO 26262 ISO 26262-10:2012 (E).

[5] SAE International. Cybersecurity guidebook for cyber-physical vehicle systems, SAE J3061 201601.

[6] W. Hoxk, C. Witteveen and M. Wooldridge, Decomposing constraint systems: equivalences and computational properties, Proc. 10th Int. Conf. of Autonomous Agents and Multiagent Systems, 2011. pp.149-156.

Self-Consistent NLOS Detection in GNSS-Multi-Constellation Based Localization under Harsh Conditions

Pierre Reisdorf, Gerd Wanielik

Technische Universität Chemnitz
Reichenhainer Str. 70, 09126, Chemnitz, Germany
Email: firstname.lastname@etit.tu-chemnitz.de

***Abstract*—Nowadays, precise and reliable localization is a key technology for many applications around the world. The localization should have an availability around the globe for a minimum price and for each person. Localization with satellites like Global Positioning System (GPS) or Glonass has become a standard in the last years for outside positioning. Satellites are available around the world and the positioning results are good for many applications, except in restricted areas. In urban areas in particular, the reception of the satellite signals is restricted. Often, the signals are received under Non-Line-Of-Sight (NLOS) conditions in these areas. Hence, a good approach should detect NLOS and use knowledge to handle the NLOS measurements. We implement a self-consistent approach to detect NLOS in the measurement domain and use the information for the position estimation process. In contrast to other existing approaches, we need no external information or hardware.**

***Keywords*–Global Navigation Satellite System (GNSS); Non-Line-Of-Sight (NLOS) detection; self-consistent; multi-constellation.**

I. Introduction

A reliable vehicle positioning is a crucial part of nearly every Intelligent Transportation System (ITS). While the importance of localization is obvious for certain applications such as navigation, fleet management, or location-based services in general, future applications, such as vehicle-to-vehicle-communications, which are solely based on absolute positioning, are currently the main driver behind the technological developments in this field. Absolute localization as implemented by Global Navigation Satellite Systems (GNSSs) seems to be a promising technological candidate to solve this task. Although many commercial vehicles are nowadays equipped with low-cost GNSS receivers, such systems provide only limited performance in dense urban areas. Especially, when buildings or foliage are acting as reflection surfaces, so called multipath errors which are caused by Non-Line-Of-Sight (NLOS) measurements, happen to appear. A brief introduction for Line-Of-Sight (LOS) and Non-Line-Of-Sight (NLOS) in urban areas is shown in Figure 1. The reception of each satellite signal depends on the position of the receiver and the satellite to each other. Satellites S_2 and S_3 are in LOS, satellite S_4 is blocked and satellite S_1 has NLOS. If such NLOS observations are not handled carefully within the localization algorithm, an unwanted bias is introduced in the final position estimate.

The paper is structured in a related work section, followed by an idea description, preliminary results and a section for next steps regarding this topic.

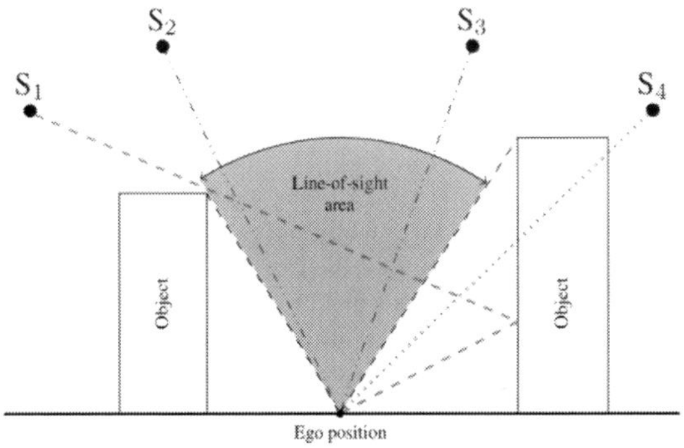

Figure 1. A typical situation in an urban area. This is a schematic representation of the electronic waves emitted by the satellites.

II. Related work

There are varying methods and approaches to detect and handle NLOS detections. A good introduction and description of different methods is given by Groves [1] and a good taxonomy is given by Obst [2]. The current approaches detect NLOS with additional hardware, special information like a 3D map with building information [3] or based on the Receiver Autonomous Integrity Monitoring (RAIM) concept with consistency checking [4]. None of these methods works with raw measurements, without additional information from external resources and can detect or exclude several satellites.

Our approach detects NLOS without additional external information and work as a self-consistent method.

III. Idea description

The NLOS effect caused by different environments generates an additional error part in the pseudorange measurement, besides the other common errors for example ionosphere, troposphere or ephemeris. A pseudorange ρ is modeled after [5] as

$$\rho = r + c(dt - dT) + d_e \quad (1)$$

$$d_e = d_{NLOS} + d_{ion} + d_{trop} + d_{eph} + \dots \quad (2)$$

In Equation (1), c is the speed of light and r the true range between the receiver and the satellite. The satellite clock error dT will transmit from the satellite by the navigation message of each satellite. The receiver clock error dt is unknown. Equation (2) describes some additional errors like ionosphere, troposphere, ephemeris or NLOS. There are various estimators and models for the different errors except NLOS. NLOS depends on the environment and the behavior is very dynamic.

The basic idea of our algorithm approach for detection of NLOS is using jumps in different parameters of raw measurements from a GNSS receiver. In more detail, the parameters are Pseudorange, Signal-To-Noise-Ratio (SNR), Carrier Phase or Doppler. A jump in this field is defined as an unusual strong increase or decrease between two or more measurements. For this purpose, we use a sliding window approach in post-processing over the last x measurements and estimate at first step the mean or median over these measurements, without the use of obvious outlier. This threshold is defined by a static long-term measurement campaign over multiple days. The second step calculates the mean or median over the complete dataset for each satellite and the result will be used in a third step as a threshold for separating as LOS or NLOS.

In post-processing, our approach used the challenging GNSS datasets from [6]. Therefore, we choose the data from the Frankfurt Main scenario for the preliminary results in Figures 2 and 3. In future work, we will use the other challenging scenarios too.

In [2], there is a good taxonomy of the multipath/NLOS problem. Our approach belongs to the class *Fault detection/ outlier classification* and open a new branch.

IV. Preliminary results

The result of the NLOS detection with the parameter pseudorange for GPS satellite 22 with the Novatel receiver is shown in Figure 2. This method to classify data by a ground truth with a Novatel receiver is described in detail in [2]. The preliminary results of the current status with the jump detection is shown in Figure 3. The results obtained with the parameters SNR and carrier phase are similar. In contrast, the doppler shows a lower rate of NLOS detection. All parameters need a deeper investigation of the different behavior with a varying size of the sliding window or other approaches to estimate the thresholds.

Figure 2 shows the detection with a state of the art method. Figure 3 depicts the state of the art method (red) and the jump approach (orange) in one diagram. Our approach detects more NLOS at the beginning of the dataset.

V. Next steps

The current approach detects NLOS measurements and excludes these measurements from the position estimation in a next step. This approach is called fault detection and exclusion (see taxonomy in [2]). In dense urban areas, the number of LOS measurements could not be enough and no position fix is available. Hence, the next step behind the classification in NLOS and LOS is to use the knowledge of NLOS data and estimate the pseudorange using this historic knowledge. A simple estimator or a filter like Kalman-filter perform this estimation process. By weighting several aspects of the jump information from the parameters, we have a chance to estimate a fixed position with the satellite information. To estimate the accuracy, we plan to implement an ego motion model and a motion model for the satellites. Accordingly, we expect a more stabilized estimation with this knowledge. Furthermore, we extend the self-consisteny approach for multi-constellation scenarios with GPS, Glonass, Galileo and Beidou.

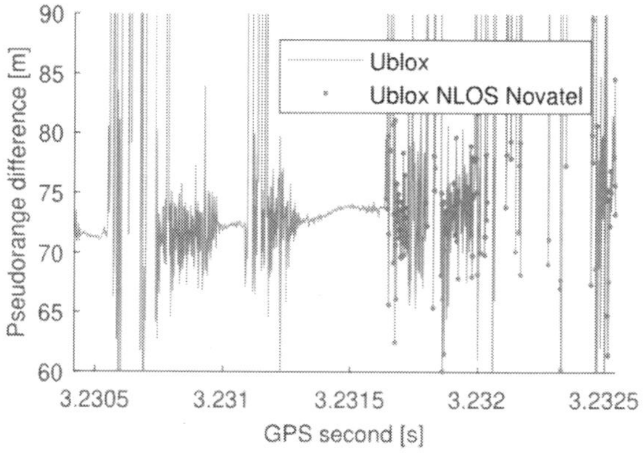

Figure 2. Ublox EVK-M8T (LEA-M8T) pseudorange measurements from GPS satellite (PRN 22) with our Novatel OEM6 NLOS (red dots) detection.

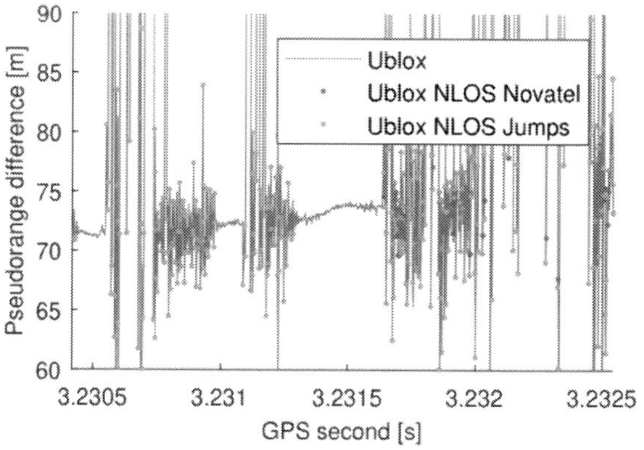

Figure 3. Ublox EVK-M8T (LEA-M8T) pseudorange measurements from GPS satellite (PRN 22) with our Novatel OEM6 NLOS (red) and our Jump NLOS (orange) detection.

Acknowledgment

For the evaluation and generation of the ground truth, precise real-time corrections provided by axio-net (http://www.axio-net.eu) were used.

References

[1] P. D. Groves, Z. Jiang, M. Rudi, and P. Strode, "A Portfolio Approach to NLOS and Multipath Mitigation in Dense Urban Areas," in ION GNSS 2013, 2013, pp. 3231 – 3247.

[2] M. Obst, "Bayesian Approach for Reliable GNSS-based Vehicle Localization in Urban Areas," Ph.D. dissertation, Technische Universität Chemnitz, 2015.

[3] M. Obst, S. Bauer, and G. Wanielik, "Urban multipath detection and mitigation with dynamic 3D maps for reliable land vehicle localization," in Position Location and Navigation Symposium (PLANS), 2012 IEEE/ION, April 2012, pp. 685 – 691.

[4] B. Parkinson and J. Spilker, Global Positioning System: Theory and Applications, ser. Global Positioning System. American Institute of Aeronautics and Astronautics, 1996, no. Bd. 1, chapter 5: Receiver Autonomous Integrity Monitoring.

[5] E. D. Kaplan and C. Hegarty, Understanding GPS : Principles and Applications, 2nd ed. Artech House, 2006.

[6] P. Reisdorf, T. Pfeifer, J. Breßler, S. Bauer, P. Weissig, S. Lange, G. Wanielik, and P. Protzel, "The problem of comparable gnss results an approach for a uniform dataset with low-cost and reference data," in The Fifth International Conference on Advances in Vehicular Systems, Technologies and Applications, M. Ullmann and K. El-Khatib, Eds., vol. 5, nov 2016, p. 8, iSSN: 2327-2058.

Reducing Car Consumption by Means of a Closed-loop Drag Control

Camila Chovet, Baptiste Plumjeaeu, Sebastien Delprat, Marc Lippert and Laurent Keirsbulck

LAMIH, Umr8201, 59313 Valenciennes, France
email: camila.chovet@univ-valenciennes.fr; baptiste.plumejeau@icloud.com; sebastien.delprat@univ-valenciennes.fr; marc.lippert@univ-valenciennes.fr; laurent.keirsbulck@univ-valenciennes.fr

Maxime Feingesicht and Andrey Polyakov

INRIA Lille
59651 Villeneuve-d'Ascq, France
email: mfeingesicht@gmail.com; andrey.polyakov@inria.fr

Jean-Pierre Richard

Centrale Lille
59651 Villeneuve-d'Ascq, France
email: jean-pierre.richard@centralelille.fr

Franck Kerhervé

Institut P', Upr3346,
86962 Poitiers, France
email: franck.kerherve@univ-poitiers.fr

***Abstract*—We present a closed-loop separation control experiment on a real car (Renault Twingo GT) using a robust model-based strategy known as Sliding Mode Control (SMC) and whose goal is to decrease the gas consumption through the reduction of the drag force. The study also investigates the feasibility of this control and estimation approach for an industrial framework. A movable spoiler, equipped with slotted air jets, is placed on the top trailing edge of the car in order to actuate the air flow on the rear side of the vehicle. In order to estimate the drag force, the vehicle is instrumented with dSPACE and DEWESoft equipment, which are able to do synchronized measurements in real-time. Concerning the control algorithm, Sliding Mode Control was designed on the basis of a non-linear delayed input-output model, which was recently defined for another flow control application and has shown a good ability for robustness and disturbance rejection.**

Keywords–Flow Control; Vehicle; Sliding Mode Control; Drag reduction.

I. Introduction

In the last years, the sales of automobiles (including cars, trucks, and buses) have sky-rocketed. More than 1 billion vehicles were in operation in 2010 [1]. The amount of transportation automobiles is highly correlated with carbon-dioxide CO_2 emission in the atmosphere, and the strong decrease of oil reserves, to name just a few consequences [2]. Car manufacturers are consequently urged to develop new technologies, which can contribute to significantly diminish the environmental impact of transportation [3]. In this regard, four development areas are identified to improve energy consumption, autonomy and gas emissions: engine performance, reduced friction of the bearings, the lightening of the mass of the vehicle and the aerodynamics [1]. The work in progress presented in this article targets the latter one. The power consumed to resist the drag constitutes an important portion of the total power expense. For example, the aerodynamic drag of a car, at a speed of 50 km/h, accounts for 50 % of the total drag, reaching 80 % at 130 km/h [4]. Initially, drag reduction was achieved through vehicle shape. This explains why most of the present cars have similar enclosures [5]; vehicles tend to follow a unique shape in order to optimize drag reduction. Nonetheless, there is room to further reduce the aerodynamic drag thanks to the manipulation and/or control of the wake flow. This approach is of great challenge for the transport industry and must be tackled as a multidisciplinary field. A combination of passive and active devices might decrease the drag force by enhancing the shear layer deflection [6]. Our aim is to tackle this industrial problematic through the knowledge of different academic areas.

The main goal is to develop an effective control strategy combined with an industrial actuator to minimize the drag of a real car and thus reduce both fuel consumption and gas emissions.

The paper is organized into three sections: Section II presents the instrumented vehicle. Section III is devoted in estimating the drag force for a real car configuration, while Section IV briefly defines the design of the closed-loop strategy on an Ahmed body configuration. It will be shown how Sliding Mode Control is able to reduce and maintain the drag to a desired set-point regardless of external flow perturbations. A short conclusion and perspectives are presented in Section V.

II. Instrumented Vehicle

The instrumented vehicle used for experiment proposes is a Renault Twingo II C44 GT owned by LAMIH laboratory. The dimensions of the car are as follows: height H of 1.47 m (from the top part of the car to the floor), width w=1.39 m and, length l=3.69 m (see Fig.1(a) and Fig.1(b)). Due to the presence of two people inside the vehicle, it has two different ground clearances. The first one closer to the front wheel g_1=38.6 cm, while the second is measured at the back wheel and is equal to g_2=51 cm. The results presented are for car velocities from 45 km/h to 130 km/h, which correspond to Reynolds numbers, based on the car height $h \approx 1.02$ ($h = H - (g_1 + g_2)/2$), of $Re_h \approx 6 \times 10^6 - 19 \times 10^6$. The results will be described in a Cartesian coordinate system x, y and z representing the stream-, transverse- (normal to ground) and span-wise directions, respectively (see Fig.1(a) and Fig.1(c)). Experiments will take place on a 2km race track

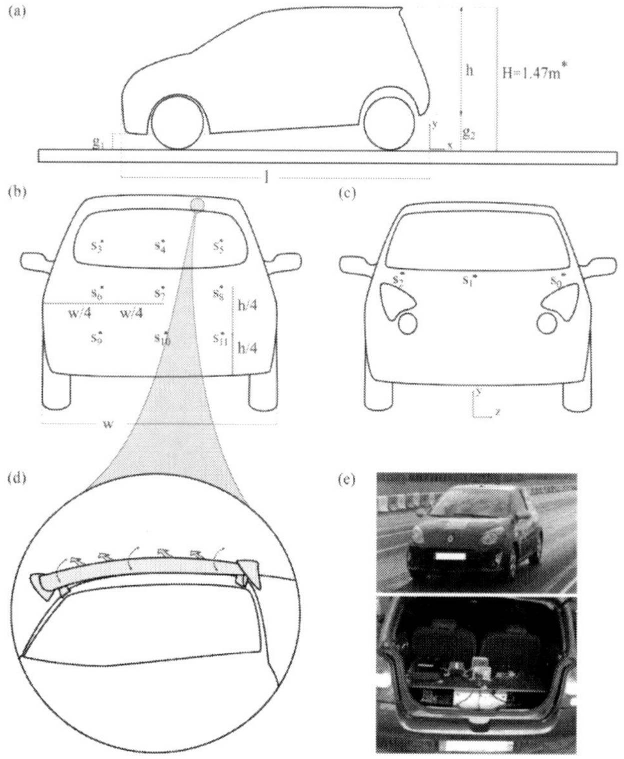

Figure 1. Sketch of the vehicle Twingo GT. (a) Rear view, (b) Front view, (c) Actuation mechanism, (d) Side view, (e) Real vehicle image.

located at Clastres (North of France). The track is travelled in both directions, direction I (dI): departure to arrival and direction II (dII): arrival to departure.

The car is equipped with a GPS, which is able to measure the longitude and latitude coordinates (x and z directions, respectively). The slope of the road is retrieved from GPS measurements. GPS data is projected on a local plane (Lambert Projection 93), and the $x-z$ data is then projected on a local coordinate system aligned with the track and centred at the end of the runway. The z-coordinate measures the lateral deviation from the reference axis. Furthermore, the car is also equipped with two acquisition systems: DEWESoft and dSPACE. Since two measurement devices are used, the data must be re-calibrated. For this, a binary signal is sent from dSPACE to DEWESoft. The registration is then carried out by synchronizing the first rising edge. Twelve sub-miniature piezo-resistive Kulite XCQ-0.62 sensors are placed at the front and the back of the car. Three of them are placed at the front close to the windshield, the first one is located at the center plane w/2 from the side of the body while the other two are at distance w/8 from the car side edge (one at each side). On the rear, 9 pressure sensors were also set. The distance between each sensor is $y = h/4$ and $x = w/4$ starting at the same $x - y$ distance, for a clear view of the sensor positions please refer to Fig.1(b) and Fig.1(c). Finally, an innovative actuation mechanism will be placed at the top trailing edge of the vehicle (see Fig.1(d)). A movable spoiler (commercially available) will be equipped with slotted jets in order to vary key actuation parameters such as angle, pressure (jet velocity), DC (Duty cycle) and frequency.

III. Force Estimation

In order to estimate the drag force, two mechanisms are here presented. The first one is with an approximation of Onorato relation [7]. 80% of the total drag force is related to the pressure forces. From this statement, the integral balance of momentum is simplified as:

$$F_x = \iint_{S_w} (Pt^{\infty} - Pt^{S_w})\, d\sigma \qquad (1)$$

where S_w denotes the surface of the wake downstream the vehicle, Pt^{∞} and Pt^{S_w} are the upstream pressure and the pressure in the wake S_w. With this equation, the drag coefficient C_x can be calculated from equation presented by Onorato [8]. Taking into account the twelve pressure probes placed in the car, and an equal area for all the sensors, the drag coefficient can be approximated as

$$C_x \approx \frac{\sum_{n=1}^{3}(Pt_n^{\infty}) - \sum_{n=1}^{9}(Pt_n^{S_w})}{1/2\rho U_{\infty}^2} \qquad (2)$$

The second one is through an estimation of the external forces presented in the car. Once a portion of the data is selected, it is now possible to estimate the external forces exerted on the vehicle. The dynamics of the vehicle is expressed in the form of an energy balance equation:

$$M\frac{dv(t)}{dt} = F_{mot}(t) \pm mgsin(\vartheta(d(t))) - F_{ext}(t) \qquad (3)$$

The longitudinal acceleration $\frac{dv(t)}{dt}$ is obtained through the central inertial unit, while the slope ϑ was previously measured using the average slope measurement as a function of the distance $d(t)$ travelled. The $\pm$ sign in equation.3 is related to the fact that the track is travelled in both directions. In the direction I, the vehicle undergoes a headwind and a slight rise. In direction II, the vehicle undergoes a wind of back and a slight descent. The estimation of the external force $F_{ext}(t)$ is trivial via (3). Additionally, the external force is decomposed as follows:

$$F_{ext}(t) = C_{rr}mgcos(\pm\vartheta) + 1/2\rho AfC_d(v(t) \pm v_{wind})^2 \qquad (4)$$

with v_{wind} the unknown wind velocity assumed constant. For simplification, $\alpha_1 = C_{rr}mgcos(\pm\vartheta)$ and $\alpha_2 = 1/2\rho AfC_d$. In (3) the term $cos(\pm\vartheta)$ is assumed constant because of the co-sinus function is even, so the sign of the slope will not influence the results. Each test is thus summarized to a pair $(\bar{v}(i), \bar{F_{ext}}(i))$. Then, the parameters are estimated by minimizing a quadratic model error:

$$J = \sum_{i.e.dI} (\bar{F_{ext}}(i) - \alpha_1 - \alpha_2(\bar{v}(i) - v_{wind})^2)^2 + \sum_{i.e.dII} (\bar{F_{ext}}(i) - \alpha_1 - \alpha_2(\bar{v}(i) + v_{wind})^2)^2$$

The minimization of J, starting from a realistic solution, relates to an optimization problem and leads to solving a linear problem for α_1 and α_2, which can be solved in a straightforward manner. The introduction of wind speed complicates the calculation of the solution and a Trust-Region Quasi-Newton algorithm is necessary.

The drag coefficient C_x is presented in Fig.2 for a velocity range of $U_{\infty} \in [45\ km/h, 130\ km/h]$ calculated from the

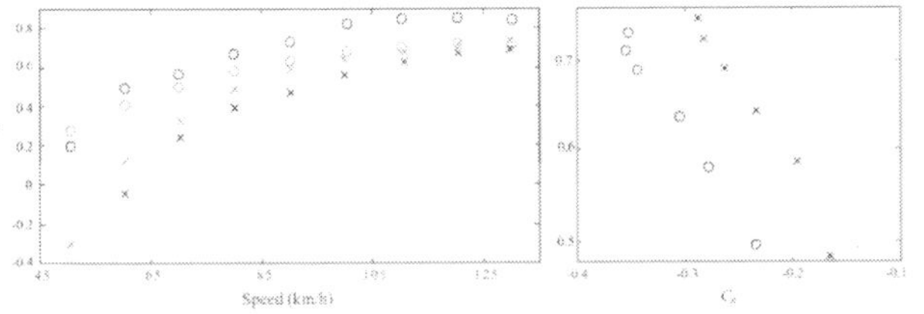

Figure 2. (a) C_x obtained for different velocities using the two approaches. (b) Relation between C_x and C_p.

two estimation approaches (red symbols: results using estimation approach, black symbols: results using Onorato's drag approximation). The figure shows both direction results. A positive curved increase of the C_x is present for an increase in the velocity. Similar results are obtained for both approaches, which means there is a good estimation of the drag force. A small gap is seen from the results in direction I (departure to arrival, circle symbols) and direction II (arrival to departure, times symbol). This is due to the effects of the wind that clearly induced some variations in the results depending on which direction the track is run. The ratio between the C_p obtained from the pressure sensor and the C_x estimated from the energy balance is presented in Fig.2(b). A linear relation is obtained, for both directions. It is possible to conclude that both methodologies could be used as sensors to estimate the drag of a real car. It is worth noting that external parameters, such as wind speed, probably varied and therefore errors are introduced in the measures.

IV. EXPERIMENTAL RESULTS OF SMC APPLICATION

As previously stated, the goal is to reduce and further maintain the aerodynamic drag of a square-back Ahmed regardless of incoming flow perturbations or again measurement noise and model inaccuracies. To achieve this, we consider the first closed-loop separation control experiment on a simplified car model, also known as Ahmed body, using a robust, model-based strategy called Sliding Mode Control. This control approach is based on an initial black-box model identification presented by Feingesicht et al. (2016 & 2017) [9][10]. Without loss of generality, the sliding mode principle is a problem of a set-point tracking where a control system is considered as the form,

$$\dot{s}(t) = f(s,t) + g(s,t)b(t) \tag{5}$$

where f and g are unknown functions and $b(t) \in \{0,1\}$ is the relay control input, s is the state and t is the time. The objective is to determine a control that guarantees $s(t) \to s^*$ as $t \to +\infty$, where s^* is a desired set-point. The control must be also robust with respect to some uncertainties in f and g. If we can force the dynamics of the system to lie on a well behaved surface, then the control problem is greatly simplified. Sliding Mode Control [11] is based on the design of an adequate "sliding surface" (or "sliding manifold") defined as,

$$\Sigma = \{(s,t) \mid \sigma(s,t) = 0\} \tag{6}$$

that divides the state space into two parts, which correspond to one of the two controls, which commutate between two pre-defined values (typically zero and one or +V and -V for a relay) when the state crosses the surface in order to maintain the sliding mode $\sigma(s(t),t) = 0$. The manifold Σ is defined in such a way that the error $|s(t) - s^*|$ vanishes to zero when the system state s is restricted to lie on this surface. The control problem then reduces to the problem of reaching phase during which trajectories starting off the manifold Σ move toward it in a finite time, followed by a sliding phase during which the motion is confined to the manifold and the dynamics of the system are represented by the reduced-order model.

An identification procedure of the dynamical system is employed here for flow control application. In the present case, the sensor output utilized is the instantaneous drag $F_D(t)$, the control variable $s(t) = F_D^{\infty} - F_D(t)$, where F_D^{∞} is a drag in the uncontrolled steady state, and the control command is the switching function $b(t)$ send to the actuators. Due to the non-linear nature of the governing equations driven the flow dynamics, non-linear models with time-delays are here preferred. The bilinear model proposed by Feingesicht et al. (2016 & 2017) [9], [10] is here considered and can be written as,

$$\dot{s}(t) = \alpha_1 s(t-h) - \alpha_2 s(t) + (\beta - \gamma s(t-h) + \gamma s(t-\tau))b(t-h) \tag{7}$$

where α_i, β, γ, h and τ are positive constant parameters to identify. Identification of the model parameters is performed thanks to open-loop control experiments, which consists in successive cycles of actuation and relaxation phases for different values of the control parameters (here frequency F_A and duty-cycle DC). For conciseness, the reader is referred Feingesicht et al. (2017) [10] for full details of the identification procedure. In the present case, it leads to $\alpha_1 = 27.37$, $\alpha_2 = 32.70$, $\beta = 1.97$, $\gamma = 1.92$, $\tau = 0.18$ s and $h = 0.01$ s. The precision of the model has been evaluated by a FIT index, presented by Chovet et al. [12] equal to 59% for the present application.

Experiments were conducted in a closed-loop wind tunnel (2 m wide, 2 m high and 10 m long). The blunt-edged bluff body is a simplified car model known as square-back Ahmed body, Fig.3(a), with the following dimensions: height h=0.135 m, width w=0.170 m and length l=0.370 m. The model was mounted over a raised floor with a ground clearance of g=0.035 m. The Reynolds number was kept constant at $Re_h = 9.10^4$. Based on previous open-loop results, the set-point $s^* = -2.5\%$ is considered as the tracked value. Drag force was measured using a 6-components DELTA-ATI aerodynamic balance. The velocity flow fields were obtained using a standard two-component TSI Particle Image Velocimetry (PIV) system. The system consists of a double-pulse laser system generating the light sheet and two cameras recording the light scattered by the tracer particles. The model is equipped with an actuator slit at the top trailing edge. The slit width is $h_s = 0.5\ mm$ and the actuation length is $w_a = 150\ mm$. The pressured air, supplied by a compressed air reservoir, can be blown tangentially to the free-stream velocity through the slit (see Fig.3(b)). The pulsed blowing is driven by a FESTO-MH2 solenoid valve that can generate an on/off pulsed jet within an actuation frequency range between $f_A = 0 - 500\ Hz$. A rounded surface, adjacent to the slit exit, with an additional plate is installed to blow the jet in a predefined direction (for this case we used an actuation an $\theta = -6°$, shown in Fig.3(c). We choose a pressure value of $P_a = 6\ bars$ and a forcing frequency control DC of 50%, giving a maximum jet velocity of 32.2 m/s. The real-time processing was achieved using an Arduino board, with a sampling rate fixed at 100 Hz.

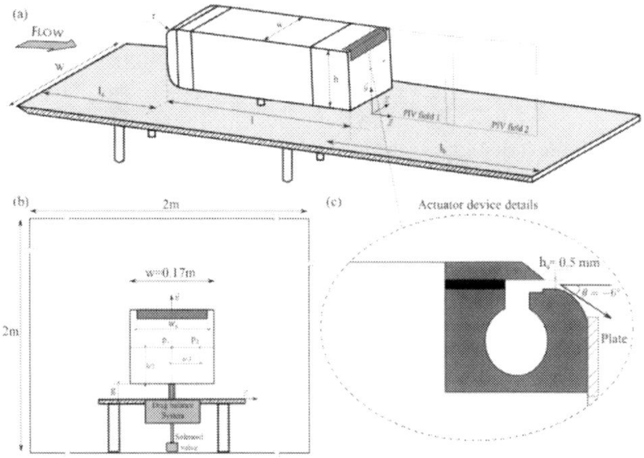

Figure 3. Ahmed body setup

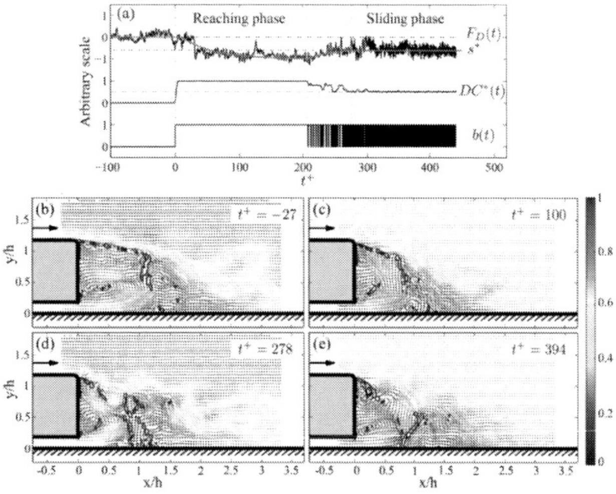

Figure 4. SMC set-point tracking results: upper plot shows the time history response. Instantaneous flow field captured by the PIV for several forced configurations.

An overview of the results obtained with the SMC is shown in Figures 4, for a further analysis of the wake dynamics please refer to [12]. Figure 4.a shows the time response of the drag (top line) for the forcing command $b(t)$ when the closed-loop controller is activated (bottom line). The calculated instantaneous duty-cycle DC^*, presented in [12], is also reported in this figure (middle line). Snapshots at four distinct instants of these time histories are reported in Figures 4.b-e. Three distinct control phases are noticeable in Figures 4.a. When the actuation is activated ($t^+ = 0$), the controller enters a phase known as "reaching phase" whose objective is to approach rapidly the targeted set-point and during which the drag force is found to decrease rapidly. In the present case, this phase corresponds to a steady blowing forcing as seen in Figures 4.a where the DC^* is found equal to unity. A transition phase ($200 < t^+ < 280$) is then observed during which the controller seeks the adequate ON-OFF combination to rapidly reach the targeted set-point. Finally, the last phase ($t^+ > 280$) corresponds to the so-called "sliding phase" where the controller attempts to maintain the drag at the tracked value s^* despite any disturbances on the plant.

V. Conclusion

The propositions presented in this short paper may be used as a reference to orient the work of flow control towards drag reduction. The ideas proposed here are a consequence of all the work done for simpler configurations. Their use in an industrial application or as an 'experimental showcase' is of great promise in future works. The car characteristics, equipment, and actuator are presented in Section II, the vehicle is already equipped with apparatus able to do measurements such as pressure, position, etc. Furthermore, a movable spoiler with jets will be added in order to control the rear wake. To achieve closed-loop control, two approaches are proposed in Section III. The first one is an approximation of Onorato's equation while the second one is an estimation of the drag force through external forces equation. Similar drag coefficients C_x were obtained for both propositions. The initial results in this experience present the possibility to directly estimate the drag of a car. Finally, we present a novel, robust closed-loop control to reduce and maintain the drag. We are interested in steering the drag of an Ahmed body to a set-point s^* and stabilizing it regardless of external conditions. In principle, the control approach can be seen as a compilation of open-loop control tests. However, the actuation frequency has to be selected carefully in real-time depending on the external parameters up- and down-stream of the flow. Otherwise, actuation energy is wasted. The robustness and easy implementation in an experimental set-up make this approach convenient for industrial applications. Also, SMC might be able to take into account model design errors of drag estimation, keeping the drag at a certain value, regardless of these errors. Hence, the benefits of SMC can be fully exploited.

Acknowledgment

This work is being held within the framework of the CNRS Research Federation on Ground Transports and Mobility, in articulation with the ELSAT2020 project supported by the European Community, the French Ministry of Higher Education and Research, the Hauts de France Regional Council. The authors gratefully acknowledge the support of these institutions.The authors declare that they have no conflict of interest.

References

[1] R. Li, "Aerodynamic drag reduction of a square-back car model using linear genetic programming and physics-based control". PhD. *thesis*, Pprime, Poitiers, France, 2017.

[2] D. Sperling and D. Gordon, "Two billion cars: transforming a culture". *TR news 259.*, 2008.

[3] W. -H. Hucho, "Aerodynamics of road vehicles". *Society of Automotive Engineers*, e.d. 1998.

[4] S. L. Brunton, and B. R. Noack, "Closed-loop turbulence control: Progress and challenges". *Appl. Mech. Rev.* 67(5), 050801, pp.01-48, 2015.

[5] G. Rossitto, "Influence of afterbody rounding on the aerodynamics of a fastback vehicle". PhD. *thesis*, École Nationale Supérieure de Mécanique et d'Aérotechnique, Poitiers, France, 2016.

[6] R. J. Englar, "Pneumatic heavy vehicle aerodynamic drag reduction, safety enhancement, and performance improvement". *The Aerodynamics of Heavy Vehicles: Trucks, Buses, and Trains*, pp. 277-302, Springer, 2004.

[7] M. Onorato, A. F. Costelli and A. Garonne , "Drag measurement through wake analysis". *SAE, SP-569, International congress and Exposition*, Detroit, MI, pp. 85-93, 1984.

[8] Y. Eulalie, "Aerodynamic analysis and drag reduction around an Ahmed bluff body". PhD. *thesis*, Bordeaux Univ, 2014.

[9] M. Feingesicht, C. Raibaudo, A. Polyakov, F. Kerhervé and J. -P. Richard, "A bilinear input-output model with state-dependent delay for separated flow control". *European Control Conference*, pp. Aalborg, Denmark, 2016.

[10] M. Feingesicht, A. Polyakov, F. Kerhervé and J. -P. Richard, "SISO model-based control of separated flows: Sliding mode and optimal control approaches". *Int. J. Robust and Nonlinear Control*, 27(18), pp. 5008-5027, 2017, DOI:10.1002/rnc.3843.

[11] V. Utkin, "Sliding Modes in Control Optimization". *CCES Springer-Verlag*, Berlin,1991.

[12] C. Chovet et al., "Sliding mode control applied to a square-back ahmed body". *J. Fluid Mech.*, 2018*b* (Under revision).

A Proposal for a Comprehensive Automotive Cybersecurity Reference Architecture

Christoph Schmittner, Martin Latzenhofer, Shaaban Abdelkader Magdy, Markus Hofer
Center for Digital Safety & Security
Austrian Institute of Technology
Vienna, Austria
Email: {christoph.schmittner | martin.latzenhofer | abdelkader.shaaban | markus.hofer}@ait.ac.at

Abstract— **Interconnection, complexity and software-dependency are prerequisites for automated driving and increase cybersecurity risks for the whole transportation system. Hence, information and communication technology infrastructure becomes a second layer for critical transportation infrastructure. In a recently started research project, we identify involved stakeholders and risks in a structured manner to integrate the diverging interests and objectives of authorities, road infrastructure providers and transport facilitators, which cannot even exclusively leave to the original equipment manufacturers. Based on the emerging risk scenarios, we develop a comprehensive architectural reference framework. Only if all components in the Information and Communication Technology (ICT) infrastructure provide their services in a sufficient quality according to ensured security requirements, society can rely on the reliable automotive system.**

Keywords- *automotive cybersecurity; road traffic infrastructure; reference architecture; risk management; ICT*

I. INTRODUCTION

Automated driving in complex and multi-modal environments for smart urban mobility requires approaches, which interconnect vehicles with other road users and infrastructure. Main benefits of connected vehicles are a reduction of accidents due to the communication of road hazards and critical situations, as well as an increase of traffic efficiency through platooning and real-time traffic monitoring and control. Reliable connectivity is the mandatory prerequisite for processing various states of the automated vehicle and accelerating further development. Positioning, a creation of complete situational awareness, reduction of accidents and increasing of comfort and efficiency depend on cooperative and automated driving. Current approaches towards stand-alone vehicles are sufficient for driving on highway or country roads, but not ready for urban environments. In addition, especially in urban environments, it is necessary to integrate automated driving vehicles into a holistic, intelligent transportation system to take advantage of all the potential benefits [1]. Therefore, this paper will focus on the infrastructure and connectivity related aspects of automated driving.

Recent projects on an European level [2] identified cybersecurity as a key challenge and risk for future transportation systems. Like physical security and protection for transportation infrastructure, cybersecurity of ICT infrastructure for connected and automated vehicles cannot be left exclusively to the private sector, as their interests and objectives differ, as well as their restricted scopes. Extensive mobility needs the cooperation of all stakeholders, i.e., automotive Original Equipment Manufacturers (OEM), infrastructure providers and road service operators, transport facilitators, end user, physical and ICT infrastructure providers, and authorities. All these actors with their different perspectives, all the components together with their relationships are considered to of a comprehensive infrastructure system requiring intense and reliable communication among these elements on different tiers not to be eavesdropped, compromised or manipulated. This makes cybersecurity a mandatory success factor for a securely and safely connected automated transportation system, which is vital for the physical transportation infrastructure and a modern society. Society can therefore rely on safe, trustworthy automotive system.

Our contribution refers to the Austrian national security research project "cybersecurity for Traffic infrastructure and road operators" (CySiVuS), which aims to tackle cybersecurity and privacy as the key challenges for cooperative traffic infrastructures and automated driving of interconnected cars. The project moves the perspective from the OEMs to traffic infrastructure providers and road service operators. The existing and future road traffic system, together with the concerning digital infrastructure is analyzed, and different autonomous driving scenarios are collected. Significant aspects require enhanced and further matured cybersecurity standards. Based on these conditions, the objective is to work out a comprehensive automotive cyber security reference architecture. It addresses all interdisciplinary interests and objectives of stakeholders and integrates existing and other technological innovations, that will be developed in the near future. This article provides a brief overview of the project's approach and highlights the urgent need for a complete reference architecture for a (cyber) secure automotive traffic infrastructure.

This paper is divided into six sections. After this introduction, we will first give a short overview of the state of the art. Section II argues that there is no sustainable structured reference architecture that supports a broad perspective on automotive cybersecurity. This is underlined by some general scenarios from a practical point of view, which we obtained from a tailored risk management process discussed in Section III. Risks should be identified, assessed and addressed through an extensive risk management approach. Based on practical use cases, we motivate the proposal of a future transportation system in Section IV. We discuss typical use case scenarios affecting the security of these automotive services from the infrastructure perspective. It is the objective of a recently started research project to develop this comprehensive automotive cybersecurity reference architecture in much more detail, the core aspects of which we introduce in Section V. The final Section VI provides conclusions and outlooks for the near future.

II. State of the Art

For automated vehicles, the Society of Automotive Engineers (SAE) J3061 [3] defines five levels, which give a framework to classify automated vehicles. Currently, mass-market available systems reach up to level three. Examples of level three are highway automation and parking assistance systems. The best-known example is Tesla's Autopilot and Parking Assistance System [4]. Even higher levels, moving towards high driving automation or even complete automation, are already in a real-world test stage [5], but not yet publicly available. While systems up to level three can rely on in-vehicle sensors and generate the world model on-demand based on local sensor data, higher levels of automation need the previous mapping to generate a world model in which the vehicle is placed via sensor data [6]. This implies that such vehicles require external input to have the latest information and react on permanent or temporary modification in the road system. This is especially important in urban environments where other localisation approaches, relying on Global Navigation Satellite System (GNSS) or road infrastructure (road markings or roadway detection) are more challenging [6].

In the United States, the National Highway Traffic Safety Administration (NHTSA) [7] currently prepares regulations, which require connectivity for active safety features for all new vehicles sold in the US starting from 2020. Such features commonly referred as cooperative active safety, require a high level of trust on outside information and communication. Safety reasons were the urgent motivation for the OEMs to establish information communication initiated by the vehicle. Security issues – which are following a different paradigm than safety-related ones – are a rather new challenge, currently addressed only by the OEMs itself. Recent hacks show that the majority of their systems lack security protection [8], [9]. Naturally, they restrict their security focus on the vehicle itself and do not follow a holistic approach, analyzing the whole infrastructure system their cars are elements. Despite first approaches, like the H.R.701 – Security and Privacy in Your (SPY) Car Study Act of 2017 [10], cybersecurity issues are still largely handled by the vehicle manufacturer simply ignoring other stakeholders. Especially when moving towards connected, intelligent and automated transportation systems, the road traffic infrastructure need to be considered in a consequently holistic way. However, briefly summarizing the legal situation in general, new regulations are evolving, but too slow promptly and substantially fragmental. In an automated driving scenario, ICT infrastructure becomes a second layer of critical transportation infrastructure. Hence, it is still in the discussion whether and how the European "Directive on Security of Network and Information Systems" also known as the NIS Directive [11] applies to the automotive sector and what the consequences for the OEMs, as well as the road infrastructure providers are in detail. This European Directive will be enforced by the end of May 2018 and seeks to ensure a high level of network and information security by improving the common security level of the provider of critical services and digital contents. The transport sector is accepted to form such a critical infrastructure and due to the increasing interoperability, connectivity aspects, communication requirements, ICT in general, and privacy issues. Hence, there is an urgent need for a full categorization and orderly development.

The vehicles require detailed data about the environment to generate a broad overview of the current situation in real time and ensure their safe movement. The integrity of the data is a prerequisite for autonomous inter-connected driving. It is evident that automated driving scenarios are not restricted to the vehicles as a stand-alone system; rather the vehicles must interact in real-time with the other components among other vehicles and in particular with the infrastructure in order to assess the current situation. Thus, interoperability is the first key requisite for efficient traffic management, co-operative functions and coordinative autonomy [12].

Connectivity between vehicles and other traffic elements is currently still in development. While almost all new premium cars already offer connectivity via Global System for Mobile Communication (GSM) to a backend system of the manufacturer [13] this is currently driven by the motivation to reduce costly recalls due to software adaptions [14] and also by the European eCall initiative. Starting with April 2018, all new vehicles sold in Europe are obliged to be able to automatically call the nearest emergency center in the case of a crash and submit position and crash-related information [2]. Applications like intelligent coordination are already tested and evaluated in real-world scenarios [15]. In such scenarios, vehicles and infrastructure need to communicate within a defined time frame and exchange information like traffic status, travel times, road conditions and road works warnings. There are higher requirements on the connectivity for the next level of cooperation and connectivity. Although there are an Intelligent Transportation System (ITS) architecture available and connectivity scenarios defined by European Telecommunications Standards Institute (ETSI) [16], it is unclear whether vehicles will possess multiple communication systems for each service provider or the communication is handled via a central data hub [17]. Different approaches of the future communication infrastructure are presented and discussed in a report of the Cooperative Intelligent Transportation System (C-ITS) platform [2]. A conclusion is that to support interoperability, stay cost-efficient, reduce the number of attack surfaces and support future applications the connectivity should follow some sort of coordinated model, considering not only the vehicle but the complete infrastructure and service value chain.

Especially in the field of cybersecurity, there are multiple signs indicating that the current state of the art cannot adequately protect the new and vital role ICT will play in transportation. Automotive cybersecurity is slowly rising to this aspect [18] triggered by research and governmental pressure [13][19][20][21]. Technical developments and industrial awareness of new challenges are followed by the development of first guidelines for tackling the issues [22]. On a higher level, the ITS infrastructure security is also a known issue which is addressed [23]. There is still ongoing discussion who will control and provide the communication infrastructure [2]. Since all mobility and the complete road transportation sector will depend on the ICT system, it is of utmost important to clarify responsibilities and to achieve a dependable balance between private and public control.

One important discussion is who controls access to the data collected by the vehicle. There are first efforts to develop processes for addressing these issues [24]. A recent survey of the German consumer organization "Stiftung Warentest" showed that almost all connectivity solutions offered by automotive OEMs have weaknesses in privacy [25]. Personal information is exchanged without encryption, and the superfluous information is collected and transmitted, partially done without informing the user and his agreement.

III. Risk Managment

There is currently no domain-specific risk management framework available for the automotive domain. First approaches [22] are promising, but initial evaluations show certain challenges in the application [26]. The guidebook [22] was published at the beginning of 2016 and after being available for half a year again set to work in progress status. The International Organization for Standardization (ISO) and SAE founded a common working group developing a standard for the cybersecurity engineering of road vehicles (ISO/SAE 21434), but the publication is currently envisioned for 2020. In the absence of applicable domain-specific frameworks, we propose to tailor ISO 31000 [27] for the application in the automotive domain. To set up the context, define the stakeholder and the application environment, an appropriate management framework has to be established first. A second main part of the risk management standard proposes the following steps of the risk management process:

1. Establishing the Context
2. Risk Assessment
3. Risk Treatment
4. Monitoring and Review
5. Communication and Consultation

Firstly, we present the framework with suggestions on how to tailor it towards the application field. The suggested tailoring will partially be done on a higher level.

A. Establishing the Context

The previous state of the art overviews shows that there is currently no specific regulatory or legal framework for road traffic. This means we can only apply generic rules and base the context on the environment and values of the society for road traffic. There are ongoing discussions about which regulations should be applied to the road traffic domain, but no clear consensus emerged so far. The automotive and transportation domain is an important part of ensuring and enabling our modern lifestyle and we, therefore, consider following objectives as necessary. It should be avoided, that a cybersecurity attack

- causes immediate damage to environment or human lives (safety);
- causes the loss of control over personal information (privacy);
- causes financial damage (finance); and
- negatively impacts the operation and traffic flow (operation).

We propose two restrictions to these objectives. Firstly, we restrict the risk management to direct and immediate consequences. This means that we do not consider second-level consequences, e.g., an operational impact would also impact emergency services and could, therefore, cause damage to human lives. Our focus lies on the direct consequences. Secondly, we assess the impact rating on users and society higher than the impact on the organization. That means that safety impacts and rate financial impacts for users or society are higher prioritized than for organizations. Society needs to trust and rely on the transportation system, which is supported by ensuring their needs and protection first.

B. Risk Assessment

Risk assessment includes identification, analysis and evaluation of risks. While ISO/IEC 31010 presents examples for risk assessment techniques, none of them is tailored for cybersecurity in the road traffic domain. There exist multiple proposals to extend established safety risk assessment methods towards cybersecurity [28], [29] or to tailor cybersecurity methods for the automotive domain [30], [31]. It should be remarked that there is no silver bullet to risk assessment implementation and any selected methodology needs to be justified. Depending on the abstraction level, different methods are favored. We propose threat modelling [33] for the analysis of risks. For risk evaluation purposes, we propose four impact levels, divided into four categories, as shown in Table I. This is an abstraction of the categories proposed by SAE J3061 [22] and EVITA [34]. Both use similar categories, but with more levels per category.

TABLE I. Impact Levels

	User / Society	Service provider / company
Safety	1	-
Operational	3	4
Privacy	2	3
Financial	3	4

A critical factor for risk evaluation in cybersecurity is the evaluation of likelihood. While for transportation domains it is discussed only to consider the impact of risk evaluation [35], this could move the focus on very unlikely risks. Details of the likelihood assessment are presented in [26], but in short, we propose to evaluate the likelihood based on the following four parameters:

- Assumed attacker capabilities
- Ease of gaining information about the systems
- Reachability and accessibility of the system
- Required equipment for an attack

C. Risk Treatment

Risk treatment is based on an assessment whether the risk is tolerable for the society, which means that the benefits of the connected and automated road traffic scenario outweigh the risk. Unless this is the case, we need to either modify the risk by implementing specific technical or organizational measures or avoid the risk altogether by deciding not to implement the scenario. Each risk treatment needs to be followed by an assessment of the effectiveness of the treatment, e.g. if the remaining risk is tolerable and can be accepted. Risk treatment assessment also includes the evaluation if the chosen measures influence other risks or scenarios

D. Monitoring and Review

There is currently no clear responsibility for monitoring and review of risks. This is impeded by the hierarchical silo structure which dominates the automotive domain at the moment. OEMs have a restricted system view and are only able to identify risks on this level. Suppliers are responsible for the implementation of risks treatments for their specific contribution and can detect change requirements. There is no unambiguous allocation of the risk monitoring responsibilities. Established approaches in the automotive domain follow mainly an incident based approach, i.e., reactive behaviour. For cybersecurity challenges, active monitoring and reaction are necessary. We propose to assign a reporting responsibility and develop a cyber incident response plan.

E. Communication and Consultation

As a continuous and parallel step along the risk assessment, treatment and monitoring, the complete management process needs to be recorded, documented and communicated to the stakeholders. This includes capturing the decisions, results and most importantly the justification for decisions and actions. Only this step makes the risk management transparent and comprehensible. It should be remarked that such records are sensitive and could be potentially misused by attackers.

IV. Practical Use case

In the CySiVuS project, we will analyze different use cases to develop the overall context requiring a comprehensive reference architecture. The first collection of use cases is based on the C-ITS Day 1 Use Case [36]. Day 1 refers to the first set of uses cases implemented and evaluated in the European Corridor – Austrian Testbed for Cooperative Systems (Eco-AT) project. One typical use case is the RoadWorks Warning (RWW) use case. This use case describes an interaction between vehicles and cooperative roadside elements, which provides information to about short time modifications in the road infrastructure to optimize traffic flow and driving strategy. Further, we analyze the Intersection Safety (ISS) use case. This use case refers to an interaction between vehicles and cooperative roadside elements, which provides information to optimize the traffic flow and driving strategy. In Eco-AT the transmitted data will only be used as information for the vehicle driver. We will consider the next step and assume that vehicles will in the future automatically act based on the received information in the future. In addition, we will also set up a third Vehicle to Vehicle (V2V) use case, e.g., a vehicle is broadcasting information about position and speed to enable other vehicles, which cannot detect the information with the vehicle sensors to consider it in their planning.

Figure 1 depicts the introduced three use cases. The RoadSide Unit (RSU) sends information to all vehicles about a temporal change in the road shape. Vehicles A and B coordinate how B, which is not visible to A, enters the main road and all vehicles receive information from the traffic light

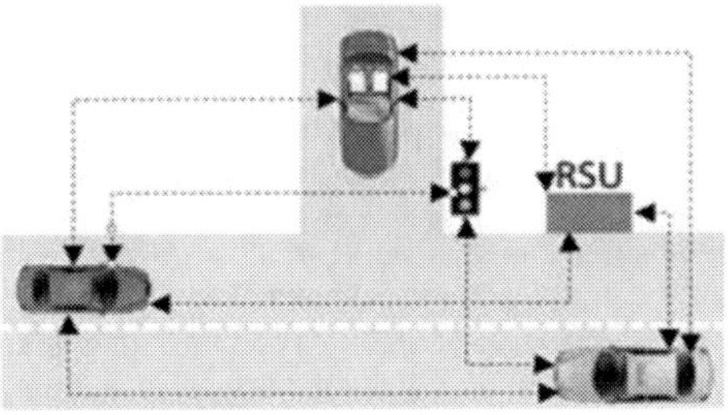

Figure 1. Use cases

system. We exercise the risk management approach based on the RWW use case. We apply threat modeling [31] for the risk assessment step.

Figure 2 visualizes the dataflow model for the interaction between vehicle A, B and the roadside units. Without any mitigation measures, twelve threats were identified. We focus in the following on the interaction type and the following threat, seen in Table II.

For connected automotive vehicles and their corresponding control and steering algorithms, the correct and especially secure reception of safety and kinematic related messages is of utmost importance. A manipulated sending unit from some distance away could communicate status information, e.g., nonexistent barriers, road works or vehicle positions ahead leading to slow down or even stop of the traffic culminating to accidents. To prevent such a threat, we propose distance-bounding protocols that allow a safe decision if the communication partner is within a certain radius, defined as bubble [37], [38]. The adapted Table III summarizes, the considerations above. This capability requires the introduction of a bidirectional communication link between Verifier (V) and Proofer (P) and a fast processing of the challenge sent from V to P. This reduces the evaluated attack likelihood by enforcing physical access to conduct such attacks and reduces the risk to a tolerable level.

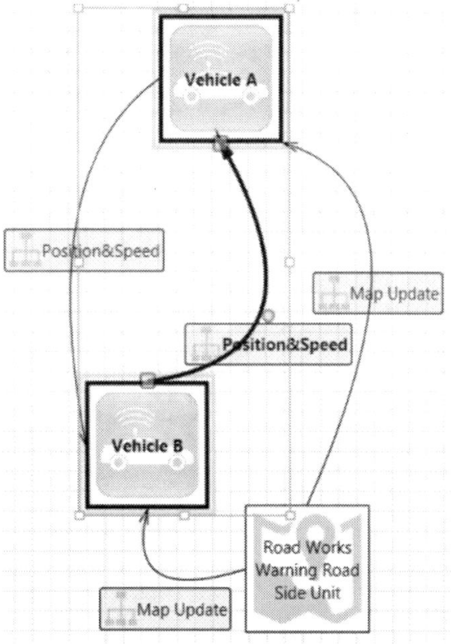

Figure 2. Data flow model for threat assessment

TABLE II. DELIVER MALICIOUS UPDATES TO VEHICLE B [PRIORITY: HIGH]

Category	Spoofing
Description	Spoofing vehicle A in order to send malicious updates
Justification	<no mitigation provided>
Attack method	Impersonate the car (clone sim or similar) and then craft the malicious update

TABLE III. DELIVER MALICIOUS UPDATES TO VEHICLE B [PRIORITY: LOW]

Category	Spoofing
Description	Spoofing vehicle A in order to send malicious updates
Justification	~~<no mitigation provided>~~ *Distance bounding avoids remote attacks and requires physical access to the environment in order to conduct the attack*
Attack method	Impersonate the car (clone sim or similar) and then craft the malicious update

V. REFERENCE ARCHITECTURE

An automotive reference architecture for security analysis was presented in [9]. While it includes the elements of communication between backend and vehicle, it does not consider all relevant scenarios for C-ITS like V2V communication. Furthermore, it only defines the technical elements and does not differentiate between environments, stakeholder, objects in the architecture a division. However, this pure technical approach is not sufficient and to apply the reference architecture in practice, this division is vital.

As a first approach, we divide the ITS into five clusters of elements as shown in Figure 3. On the physical side (blue, left side), we have vehicles, infrastructure and personal devices. The provider's side (green, right side) contains elements which are maintained and operated by infrastructure operators and road service providers offering mobility services (grey, lower side) available to the users (yellow, upper side). All elements are interconnected by a communication system (orange, in the middle). It should be highlighted that these blocks can overlap, e.g., infrastructure providers can also provide services; and blocks can contain multiple diverse sub-blocks, e.g., communication collects a multitude of techniques like wireless networking (WiFi) or GSM, which can be applied for V2V or Vehicle to Infrastructure (V2I) communication.

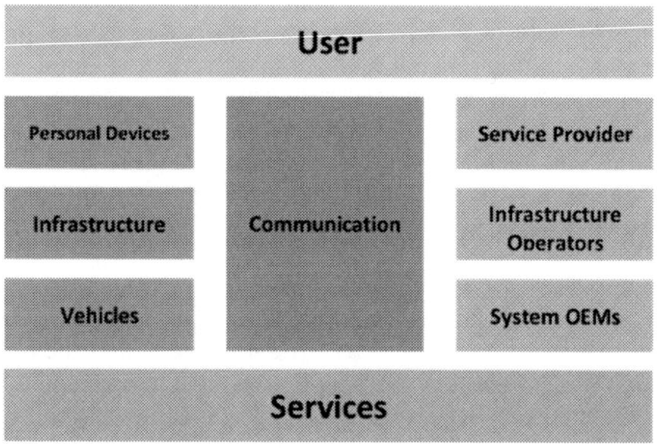

Figure 3. Clustering of elements in the transportation system

Moreover, the approach described above offers a relatively high-level view on the system, which is, to a certain degree, architecture independent. As it is discussed in [2] and [17], it is still in discussion how the connectivity architecture will finally look like, but all discussed architectural variants fit in the presented structural model. Such a structural model helps to identify the involved parties, allows assigning risk mitigations to technical elements and assigns the responsibility of implementing and maintaining these risk mitigations to involved parties. To be practically applicable, the identified risk mitigation measure is implemented in infrastructure and vehicle, conducted by system OEM and infrastructure providers, which is shown in Figure 4.

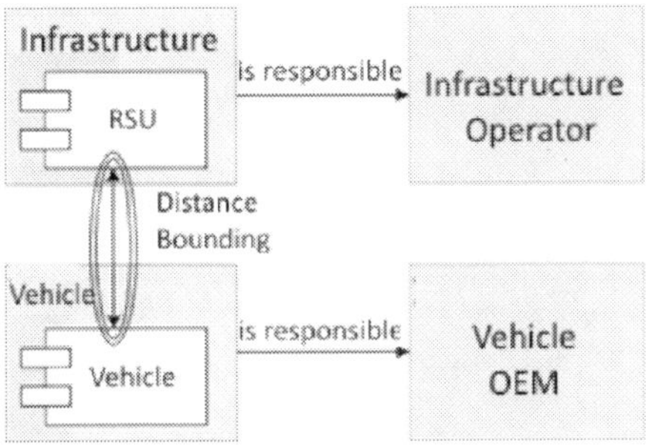

Figure 4. Application of the structural model

A possible solution approach is a structured multi-tiered reference architecture. However, a consistent risk management method is a critical success factor that this consistent architecture can be developed along all perspective layers. Our approach is to take the widely accepted risk management standard ISO 31000 [27] as a basis and tailor it to the automotive requirements. We discuss the five main steps of the risk management process when we apply it to a road traffic system. It is crucial to restrict the proposed approach to direct risks only and to weight the impacts differently depending on the consequences. The risk management analysis steps are essential to finding an appropriate mixture of applicable methods to form a reliable methodology for the assessment. Additionally, the evaluation of the likelihood and the handling of uncertainty needs to be solved. Risk treatment in a complex and interconnected environment must consider different actors

VI. Conclusion and Outlook

To conclude our contribution, the technological and legal state of the art of automated driving for smart urban mobility is still not yet sufficient to cope with the complex requirements of such an environment. We identified four current challenges in a comprehensive traffic road system. Interoperability of the components among the vehicles as well as the infrastructure elements; connectivity and communication tasks especially for interacting and cooperation of the different components; ICT in general and cybersecurity issues to address security threats; and privacy aspects which subsume protection requirements of personal data of the vehicle drivers. There are efforts to form a compliant legal and technological framework, but all these considerations are in a flow.

Finally, we discuss some previous works and propose core considerations on a comprehensive automotive reference architecture. We identify five element clusters required to interact with each other. The primary task of the CySiVuS research project is to develop a wide-ranging model on all necessary perspective levels, which the rough approach introduced in this article could be a starting point. By conducting the risk management process and developing the reference architecture, we show the multidimensional nature of a road traffic system. The main task in the upcoming period is to cope with the complexity and streamline the extremely different current and future developments on the various perspective levels.

Acknowledgment

The research project "Cybersicherheit für Verkehrsinfrastruktur- und Straßenbetreiber" (CySiVuS, in English: „Cyber security for transport infrastructure and road operators") (Project-Nr. 865081) is supported and partially funded by the Austrian National Security Research Program KIRAS (Federal Ministry for Transport, Innovation and Technology (BMVIT) and Austrian Research Promotion Agency (FFG) 2017).

References

[1] Q. Xu, K. Hedrick, R. Sengupta, and J. VanderWerf, "Effects of vehicle-vehicle/roadside-vehicle communication on adaptive cruise controlled highway systems," in *Vehicular Technology Conference, 2002. Proceedings. VTC 2002-Fall. 2002 IEEE 56th*, 2002, vol. 2, pp. 1249–1253 [Online]. Available: http://ieeexplore.ieee.org/ xpls/abs_all.jsp?arnumber=1040805. [Accessed: 27-Oct-2014]

[2] C-ITS Platform, "Working Group 6 - Access to in-vehicle resources and data," 2015.

[3] SAE, "J3016 Taxonomy and Definitions for Terms Related to Driving Automation Systems for On-Road Motor Vehicles," 2016.

[4] M. Dikmen and C. M. Burns, "Autonomous Driving in the Real World: Experiences with Tesla Autopilot and Summon," 2016, pp. 225–228 [Online]. Available: http://dl.acm.org/citation.cfm?doid=3003715.3005465. [Accessed: 27-Sep-2017]

[5] M. Aeberhard *et al.*, "Experience, Results and Lessons Learned from Automated Driving on Germany's Highways," *IEEE Intelligent Transportation Systems Magazine*, vol. 7, no. 1, pp. 42–57, 2015.

[6] G. Bresson, Z. Alsayed, L. Yu, and S. Glaser, "Simultaneous Localization and Mapping: A Survey of Current Trends in Autonomous Driving," *IEEE Transactions on Intelligent Vehicles*, pp. 1–1, 2017.

[7] United States Department of Transportation, "NHTSA | National Highway Traffic Safety Administration," 2017. [Online]. Available: https://www.nhtsa.gov/. [Accessed: 28-Sep-2017]

[8] A. Greenberg, "The Jeep Hackers Are Back to Prove Car Hacking Can Get Much Worse | WIRED," 08-Jan-2016. [Online]. Available: https://www.wired.com/2016/08/jeep-hackers-return-high-speed-steering-acceleration-hacks/. [Accessed: 29-Sep-2017]

[9] J. Brückmann, T. Madl, and H.-J. Hof, "An Analysis of Automotive Security Based on a Reference Model for Automotive Cyber Systems," presented at the SECURWARE 2017: The Eleventh International Conference on Emerging Security Information, Systems and Technologies, Rome, 2017, pp. 136–141.

[10] Library of Congress, "H.R.701 - 115th Congress (2017-2018): SPY Car Study Act of 2017," *Congress.gov*, 31-Jan-2017. [Online]. Available: https://www.congress.gov/bill/115th-congress/house-bill/701. [Accessed: 28-Sep-2017]

[11] European Union, *Directive (EU) 2016/1148 of the European Parliament and of the Council of 6 July 2016 concerning measures for a high common level of security of network and information systems across the Union*, vol. L194. 2016 [Online]. Available: http://eur-lex.europa.eu/legal-content/EN/TXT/PDF/?uri=OJ:L:2016:194:FULL&from=EN. [Accessed: 22-Aug-2017]

[12] J. Harding *et al.*, "Vehicle-to-Vehicle Communications: Readiness of V2V Technology for Application." National Highway Traffic Safety Administration, Washington DC, Aug-2014 [Online]. Available: https://www.nhtsa.gov/sites/nhtsa.dot.gov/files/readiness-of-v2v-technology-for-application-812014.pdf. [Accessed: 02-Oct-2017]

[13] C. Miller and C. Valasek, "A Survey of Remote Automotive Attack Surfaces," White Paper, 2014.

[14] H. A. Odat and S. Ganesan, "Firmware over the air for automotive, Fotamotive," in *Electro/Information Technology (EIT), 2014 IEEE International Conference on*, 2014, pp. 130–139 [Online]. Available: http://ieeexplore.ieee.org/xpls/abs_all.jsp?arnumber=6871751. [Accessed: 17-Dec-2014]

[15] Autobahnen- und Schnellstraßen-Finanzierungs-Aktiengesellschaft (ASFINAG, in English: "Autobahn and high way financing stock corporation") Maut Service GmbH, "ECo-AT The Austrian contribution to the Cooperative ITS Corridor," 2017. [Online]. Available: http://eco-at.info/. [Accessed: 24-Jan-2017]

[16] European Telecommunications Standards Institute, "Intelligent Transport Systems (ITS); Cooperative ITS (C-ITS); Release 1." 2013.

[17] B. Datler, M. Harrer, M. Jandrisits, and S. Ruehrup, "A Road Operator's View on Cloud-based ITS – Requirements and Cooperation Models," presented at the 23rd ITS World Congress, Melbourne, Australia, 2016.

[18] L. Apvrille *et al.*, "Secure automotive on-board electronics network architecture," in *FISITA 2010 world automotive congress, Budapest, Hungary*, 2010, vol. 8 [Online]. Available: http://www.eurecom.fr/fr/publication/3132/download/rs-publi-3132.pdf. [Accessed: 29-Jan-2017]

[19] D. Spaar, "Car, open yourself! Vulnerabilities in BMW's ConnectedDrive," c't, no. 5, pp. 86–90, 2015.

[20] J. Petit and S. E. Shladover, "Potential Cyberattacks on Automated Vehicles," *IEEE Transactions on Intelligent Transportation Systems*, pp. 1–11, 2014.

[21] C. Miller and C. Valasek, "Remote Exploitation of an Unaltered Passenger Vehicle," Black Hat 2015, Aug. 2015.

[22] Society of Automotive Engineers, *Cybersecurity Guidebook for Cyber-Physical Vehicle Systems (work in progress)*. 2017.

[23] ECo-AT, "ECo-AT SWP3.4 Security, Release 3.6." 2016 [Online]. Available: http://www.eco-at.info/system-spezifikationen.html

[24] International Organization for Standardization, "ISO/DIS 20077-1 Road Vehicles - Extended vehicle (ExVe) methodology - Part 1: General information," 2016. [Online]. Available: http://www.iso.org/iso/home/store/catalogue_ics/. [Accessed: 27-Jan-2017]

[25] Stiftung Warentest, "Connected Cars: Apps of the automobile manufacturer are data sniffers" ["Connected Cars: Die Apps der Autohersteller sind Datenschnüffler"], 26-Sep-2017 [Online]. Available: https://www.test.de/Connected-Cars-Die-Apps-der-Autohersteller-sind-Datenschnueffler-5231839-5231843/

[26] C. Schmittner, Z. Ma, C. Reyes, O. Dillinger, and P. Puschner, "Using SAE J3061 for Automotive Security Requirement Engineering," in *Computer Safety, Reliability, and Security*, vol. 9923, A. Skavhaug, J. Guiochet, E. Schoitsch, and F. Bitsch, Eds. Cham: Springer International Publishing, 2016, pp. 157–170 [Online]. Available: http://link.springer.com/10.1007/978-3-319-45480-1_13. [Accessed: 28-Sep-2017]

[27] International Organization for Standardization, Ed., *ISO 31000:2009 Risk management - Principles and guidelines*. ISO, Geneva, Switzerland, 2009.

[28] G. Macher, Harald Sporer, Reinhard Berlach, Eric Armengaud, and Christian Kreiner, "SAHARA: A Security-Aware Hazard and Risk Analysis Method," in *Proceedings of the 2015 Design, Automation & Test in Europe Conference & Exhibition*, 2015, pp. 621–624.

[29] C. Schmittner, T. Gruber, P. Puschner, and E. Schoitsch, "Security application of failure mode and effect analysis (FMEA)," in *Computer Safety, Reliability, and Security*, Springer, 2014, pp. 310–325 [Online]. Available: http://link.springer.com/chapter/10.1007/ 978-3-319-10506-2_21. [Accessed: 04-Nov-2014]

[30] C. Schmittner, Z. Ma, E. Schoitsch, and T. Gruber, "A Case Study of FMVEA and CHASSIS as Safety and Security Co-Analysis Method for Automotive Cyber-Physical Systems," 2015, pp. 69–80 [Online]. Available: http://dl.acm.org/citation.cfm?doid=2732198. 2732204. [Accessed: 29-Sep-2017]

[31] M. Zhendong and C. Schmittner, "Threat Modeling for Automotive Security Analysis," presented at the SecTech 2016, Jeju, 2016.

[32] G. Sindre and A. L. Opdahl, "Eliciting security requirements with misuse cases," *Requirements Engineering*, vol. 10, no. 1, pp. 34–44, Jan. 2005.

[33] Frank Swiderski and Window Snyder, *Threat Modeling*. Microsoft Press Redmond, 2004.

[34] Fraunhofer Institute for Secure Information Technology, "EVITA Project Summary," Deliverable D0, 2013 [Online]. Available: http://www.evita-project.org/Publications/EVITAD0.pdf. [Accessed: 31-Oct-2014]

[35] J. Braband, "Towards an IT Security Framework for Railway Automation," presented at the Embedded Real Time Software and Systems, Toulouse, 2014 [Online]. Available: http://www.erts2014.org/site/0r4uxe94/fichier/erts2014_7c3.pdf. [Accessed: 22-Oct-2014]

[36] Federal Ministry of Transport, Innovation and Technology, "C-ITS Strategy Austria" ["C-ITS Strategy Austria."], Bmvit, 2016.

[37] K. B. Rasmussen and S. Capkun, "Realization of RF Distance Bounding.," in *USENIX Security Symposium*, 2010, pp. 389–402.

[38] G. P. Hancke and M. G. Kuhn, "Attacks on time-of-flight distance bounding channels," in *Proceedings of the first ACM conference on Wireless network security*, 2008, pp. 194–202.

A TCO Analysis Tool based on Constraint Systems for City Logistics

Johannes Kretzschmar, Mirko Johlke, Wilhelm Rossak
Institute for Mathematics and Computer Science
Friedrich Schiller University Jena, Germany,
Email: [johannes.kretzschmar, mirko.johlke, wilhelm.rossak]@uni-jena.de

Abstract—**This paper presents an approach to extend existing total-cost-of-ownership (TCO) models by a logic and numeric constraint system. Instead of an only vehicle-centered cost view, these constraints enable the modeling of various dependencies occurring in a complex application field like logistics. The resulting tool will ensure an economical disposition, fleet management and business decision-making.**

Keywords–TCO; Constraint Model; Electrification; Logistics.

I. Introduction

There have been a lot of changes in the field of city logistics over the last years, like the electrification of vehicles (EV), new means of transport built for last-mile-deliveries or whole approaches like distribution points (hubs), car-sharing or multi-use concepts. But, despite extensive funding in research and development in this area, the transportation market is remaining reserved regarding changes. The cost analysis in [1] and [2] for example have shown, that electric vehicles only get profitable with a relatively high mileage compared to conventional vehicles. This fact is besides the hesitant development of a charging infrastructure and the limited range a primary explanation of the hesitant commitment to EVs in commercial applications. This paper introduces a conceptual method not only to support and ensure the profitability of fleet restructuring, but also to enable long-term economical monitoring, extrapolation of new scenarios and strategy development. This method is based on the idea of TCO models, as introduced in the related work section.

The rest of the paper is structured as follows. In section 2, we embed our approach in the context of current work and fundamentals. In the following main section, we will introduce the pivotal idea of the combined use of numerical and logical constraints for achieving a more comprehensive TCO analysis in logistics. We close this paper with a concept architecture for a TCO tool, as well as a summery of the problem area we will prospectively work in.

II. Related Work

A vehicle TCO model consists of all expenses which arise by the acquisition and operation of that vehicle. Thereby, it is possible to offset these costs against profits or to compare different vehicle alternatives or rather multiple deployment scenarios. Especially market changes in automotive industries over the last years and the upcoming engines based on renewable energies made a TCO approach very useful for economical decision-making. Most TCO models, like [1] or [2], are very vehicle-centered and only contain vehicle specific influencing factors like acquisition, insurance, workshop or fuel expenses. To fully implement such a TCO model into a commercial application, it is necessary to enrich the model with application specific factors. The method of this paper is based on a TCO model especially built for logistics by [3]. Besides vehicle costs, this model contains expense data of the operational context. This covers personnel costs of the driver, possible trailers or hubs and information about the cargo and customers. All TCO models have in common that the use is intended to calculate costs for a specific set of parameters. An optimization or variation is only possible by experimentally varying the parameters.

III. Constraint System

The evaluation of vehicles in logistics on an operational level does not only result from vehicle cost factors. Different means of transport may have varying requirements regarding driver qualification, cargo size and weight or temporal restrictions of customers. The choice of a vehicle is strongly entangled with a tour specification and available resources.

TCO models may be an appropriate basis for one-time decisions like the acquisition of a vehicle for a generic use case. The method of this paper though proposes a much more holistic approach by implementing the TCO model into an arithmetic constraint system and extending it by adding logic constraints. A constraint system features a set of constraints, which are processable by an automatic solver. This solver calculates a solution space for every variable in the constraint system. By doing so, the TCO model gets much more flexible. Trivially, the solver can compute TCOs, but also possible combinations of variable input factors like a specific vehicle for given TCO for example. By adding comprehensive restrictions and dependencies of a logistic domain, the system solves the satisfiability problem which is identical to a resource allocation task. A constraint based TCO model is therefore a much more valuable tool for disposition, fleet management and business strategy finding. This method differentiates two types of constraints:

A. *Numerical Constraints*

Constraints are typically over a specific domain, but can be mixed and calculated within a solver. The logistics TCO model by [3] is a set of arithmetic formulas in Matlab, which are transferred into a domain for real numbers. Depending on the particular application or nature of constraints, a finite, rational or linear domain would also be conceivable. We propose the use of CLP(Q,R)[4] library for SICStus Prolog, because this solver supports multiple types of domains. By this, the equation for overall-costs of a vehicle assigned to a tour for example

$$\mathbb{K} = (12 * K_{\text{salary}}) + (S_{\text{workdays}} * K_{\text{tour}}) + K_{\text{insurance}} + K_{\text{inspection}}$$

is transformed into an element of the constraint set in Prolog

```
clpr:{K #= (12 * Salary) + (Workdays * TourCosts) +
    Insurance + Inspection.}
```

where the `clpr` identifies the domain of the various variables.

B. Logical Constraints

A holistic approach to a cost model in logistics comprises much more dependencies though. Especially the relation between vehicle, tour and driver implies non-arithmetic constraints, like qualifications or legal restrictions. We are approaching these problems by modeling ontological constraints in a logic domain as usually done by default in Prolog. In [5], we published an ontology specialized for inner-city logistics including the relation between various driver license classes in the European Union. We hereby used this description to build a logical constraint knowledge base in Prolog. First, we defined taxonomy facts about driver licenses and their associations with drivers. As shown in Listing 1, we further formalized generic rules about how the property of a license is subsumed.

```
driverL(dlA).
driverL(dlB).
subL(dlA,dlB).

staff(bob).
hasLicense(bob,dlB).

hasLicense(X,SubL) :- staff(X), driverL(SubL),
        subL(SubL,DlX), hasLicense(X,DlX).
```

Listing 1. Basic rules for a driver license taxonomy in Prolog

This basic set of facts allows the solver to determine drivers with a specific or implied subsumed license. In combination with the arithmetic TCO model, the solver then only calculates costs for appropriate tour-vehicle-driver-allocations. The TCO gets calculated much more on purpose with a resource allocation as by-product.

IV. User and Data Interface

Because the handling of a descriptive constraint knowledge base is not feasible for average consumer, we significantly focus on the integration of a suitable graphical user interface (GUI). Figure 1 shows the overall architecture concept of our tool. The Prolog knowledge base initially consists of the terminological knowledge and basic constraints. A data importer collects data from the logistics software systems and translates these into either arithmetic or logical assertional knowledge constraints. On the user side, an interface interprets functions to Prolog querys and proccesses results into changes of the GUI.

V. Conclusion and Future Work

In contrast to the existing vehicle-centered TCO models, our method furthermore represents a holistic economical view on the disposition process. Besides the calculation of vehicle, tour or personnel costs it ensures the profitability of resource allocation by implementing the constraint system as an optimizer. By using fuzzy ranges or plain variables as TCO parameters, the tool is able to find solution spaces which is useful for decision-making by logistics experts. The holistic and descriptive model covers also a daily dispatchers, as well as long-term use for business strategists.

To date, we implemented basic concepts and rules in addition to the existing TCO model from [3] to show the feasibility. We want to extend this basic knowledge base gradually by all relevant facts and dependencies of a logistics domain. Eventually, this tool will be tested and evaluated within a field test by various newspaper and media logistic companies.

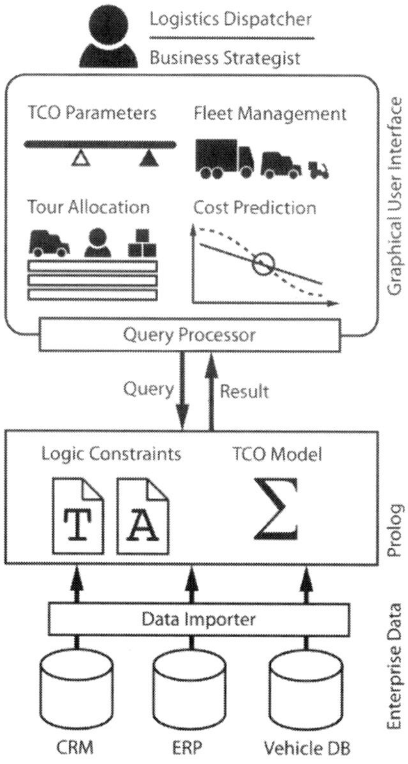

Figure 1. Conceptual Architecture of TCO-Tool

Acknowledgment

The underlying research project of this paper, Smart Distribution Logistik is funded by the german Federal Ministry of Economics and Technology (Bundesministerium fr Wirtschaft und Technologie - BMWi) during the term May 2017 to March 2020.

References

[1] G. Wu, A. Inderbitzin, and C. Bening, "Total cost of ownership of electric vehicles compared to conventional vehicles: A probabilistic analysis and projection across market segments," Energy Policy, vol. 80, 2015, pp. 196 – 214, retrieved March 2018. [Online]. Available: http://www.sciencedirect.com/science/article/pii/S0301421515000671

[2] J. Hagman, S. Ritzn, J. J. Stier, and Y. Susilo, "Total cost of ownership and its potential implications for battery electric vehicle diffusion," Research in Transportation Business and Management, vol. 18, 2016, pp. 11 – 17, retrieved March 2018. [Online]. Available: http://www.sciencedirect.com/science/article/pii/S2210539516000043

[3] A. Apfelstädt and M. Gather, "New design of a truck load network," in Dynamics in Logistics, H. Kotzab, J. Pannek, and K.-D. Thoben, Eds. Cham: Springer International Publishing, 2016, pp. 183–191.

[4] C. Holzbaur, "Ofai clp (q, r) manual," edition 1.3. 3. Technical Report TR-95-09, Austrian Research Institute for Artificial Intelligence, Vienna, Tech. Rep., 1995.

[5] T. Prinz, J. Kretzschmar, and V. Schau, "A knowledge base for electric vehicles in inner-city logistics," in The Tenth International Conference on Software Engineering Advances (ICSEA), 2015, pp. 257 – 260.

UAVs-assisted Data Collection in Vehicular Network

Mohamed Ben Brahim*†, Hakim Ghazzai*, Hamid Menouar*, Fethi Filali*
*Qatar Mobility Innovations Center (QMIC), Qatar University, Doha, Qatar
†HANA Research Lab, University of Manouba, Manouba, Tunisia
Email: {mohamedb, hakimg, hamidm, filali}@qmic.com

Abstract—**In vehicular network, nodes generate and transmit timely measured data by embedded sensors. Some data needs to be gathered by remote devices. The limited communication range of the vehicular wireless system requires to proceed through multi-hop data routing to collect periodic fresh data. The hop-by-hop based data journey and the dynamic topology increase the overall packet delivery delay, as well as the packet loss ratio. Unmanned Aerial Vehicles (UAVs) could be used in relaying vehicular data, particularly for sparse networks. Once they join the network, they are considered as normal network nodes however with some specific link characteristics. In this paper, a link-aware data collection approach is investigated. More precisely, an optimization problem maximizing a weighted multi-objective utility including the wireless link data rate, the wireless link stability, and the data progress towards the destination is formulated and solved using a Distributed Minimum Spanning Forest (DMSF) approach. The outcomes of the proposed approach and the impact of the UAVs assistance are evaluated. The present approach outperforms the other algorithms and the usage UAVs enhances the DMSF-based solution especially for low-density network.**

Keywords–Unmanned aerial vehicles; Vehicular network; Data collection; Distributed minimum spanning forest.

I. INTRODUCTION

Vehicular networks are promoting a wide range of applications rendering the driving experience safer and more efficient. Indeed, using embedded communication capabilities, the vehicles are able to exchange timely and accurate information reflecting their status and mobility. Neighbor vehicles use the received information to mitigate collisions [1] and avoid road hazards [2]. Vehicles may disseminate traffic-related information in the relevant traffic stream to early warn the incoming travelers to seek alternative routes. An Intelligent Transport System (ITS) station is usually equipped with an On-Board Unit (OBU) which is an embedded computer interfacing with a multitude of measurement-units and sensors mounted in the vehicle, e.g., Global Positioning System (GPS), proximity ultra-sound and short-range radars, etc.

Collecting data in real-time in mobile network is challenging. Unlike data exchange or data dissemination, data collection's purpose is to route data towards a gathering node from all the network nodes. The achievement of this task should take into account the underlying network characteristics. Indeed, the vehicles are moving with relatively high-speed compared to legacy mobile ad hoc network (MANET). In addition, obstructing objects and signal attenuation due to mobility and radio interference impose a wise use of the network resources. Moreover, the height of vehicle-embedded antennas and the frequency bands used in vehicular network differentiate vehicular networks from cellular networks with infrastructure support [3]. These constraints require an optimized design of communication protocols to transfer data between mobile nodes and to ensure reliable data gathering by remote infrastructure nodes.

The UAVs have recently seen an increasing usage in civilian and military applications [4]. This trend is explained by the ease of deployment and the low cost of maintenance. Among the promising applications, the UAV may contribute in the monitoring and assistance of the traffic in ITS [5]. Indeed, the drone could be equipped with wireless communication system and participates in relaying traffic data or generating traffic alerts to warn ground road users, as illustrated in Figure 1. Worthy to mention here that the energy-related issues of the drones are intentionally skipped and the aerial nodes are considered as simply hovering over the road junction at a fixed altitude. the mobility of UAVs is planned to be studied in a future work.

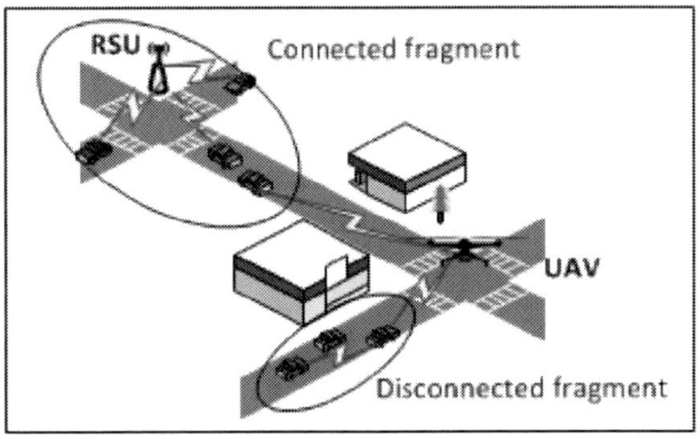

Figure 1. UAV relays data between two disconnected network fragments.

The present work proposes a new data collection approach using Distributed Minimum Spanning Forest (DMSF) algorithm: a modified version of the typical Distributed Minimum Spanning Tree (DMST) algorithm [6]. The proposed approach is based on two phases: the paths construction, and the effective data collection:

• The distributed routes construction phase proceeds with a parallel building of multiple trees having a minimum overall utility. These multiple trees enable the establishment of the routing paths and balance the load of the data traffic over the network. The utility considered in the construction of the DMSF is a weighted multi-objective function achieving a trade-off between three metrics: i) the achieved data rate over the wireless link, ii) the inter-vehicles link stability, and iii) the progress towards the destination. Hence, each node selects the relay node providing the best throughput for rapid data transfer, the highest link stability to avoid link disappearance mainly during big size packet transmission, and the closest route to the destination to reduce delays due to processing at the relay levels.

• Once the data collection session is due, nodes transmit the received data from children nodes and their data content to their parent nodes in the tree-routing structures. The process is executed by each node once it has a data to send or to forward. The data is forwarded towards the roots of the trees. Since the roots are one-hop neighbors of the sink node, a simple data forwarding ensures the successful delivery of data. When all the data is collected by the sink or the collection session is expired, the non-delivered data is discarded and a new routing topology is calculated to start a new data collection round.

The proposed approach seeks to ensure periodic data collection generated by a big number of mobile nodes through a fast and distributed routes construction. This kind of periodic data gathering is required especially for network monitoring and analysis purposes. The performances of the proposed approach are evaluated against existing algorithms.

The remainder of this paper is organized as follows: Section II provides an overview of the existing work in data collection. Section III presents the system model. Section IV starts by presenting the methodology followed by the paper. Afterwards, it presents the problem formulation. Section V introduces the proposed DMSF algorithm to solve the data collection problem. The next section presents the simulation environment and some selected numerical results. Finally, Section VII concludes the paper.

II. Related Work

For many use cases, we need to collect data from mobile nodes within the network by a collection host, aka sink. At the collection time, the traveling vehicles may or may not be close-enough to the collection node in order to deliver their data. Hence, several studies tackled this issue and investigated the possible ways to ensure data handing to the collection station. In [7], the authors presented a survey of the position-based data routing in vehicular networks. The position information availability and accuracy is a basic factor of the algorithm performances. The forwarding decision, which is usually greedy or an improved form of greedy, ensures a progress towards the destination node. Although judged to outperform topology-based algorithms, position-based routing algorithms present some limitations. Authors conclude that there is not a position-based protocol performing well in both urban and highway environments and compare the performance of the protocols mainly in terms of overhead, availability, resilience, and latency. In [8], authors leveraged a hybrid communication scheme using 3GPP/LTE and the IEEE 802.11p-based technologies to disseminate safety messages in Vehicular Ad hoc NETwork (VANET). The approach is based on the construction of multi-hop clusters and the offload of data transfer between farther clusters to the cellular network. The hybrid communication scheme outperforms the basic clustering and the flooding-based forwarding algorithms. Even though it resolves the hole problem especially in sparse networks, the proposed approach could not be the best option in the present scenario because no alternative communication technology other than the Dedicated Short Range Communication (DSRC) wireless technology is available.

Another approach for data collection based on a mobile agent is proposed in [9]. The proposed solution takes into account the lossy nature of the network links and uses mobile agents in the network to collect data from vehicles in one-hop fashion. The approach reaches a higher data collection ratio but is still limited in terms of data freshness, which is ensured by the present approach, and may be useful for special cases of deferred offline data processing scenarios. An earlier published algorithm, named ADOPEL [10], is proposed using the reinforcement learning technique for the data collection in VANET. The approach uses a distributed Q-learning technique to update Q-values required to select the relaying nodes to forward data towards the collection node. The algorithm is fully distributed and outperforms the non-learning approach. However, the fast topology updates negatively affect the convergence of the learning strategies. Indeed, the required time to learn the best relay is likely to exceed the link duration of all or some of the available links at a given time. In this paper, we overcome this issue in DMSF through periodic assessment and re-weighting of the wireless links. In [11], a data routing optimization approach based on multi-objective metrics and minimum spanning tree (MST) is proposed. The solution investigates an interference-aware routing scheme by fostering the better links and radio channels for routing the data. The used metrics to compute the link weight are end-to-end delay, link duration probability, and co-channel interference. The algorithm needs a global awareness of the network topology (e.g., convergence of routing tables) by every node to be able to compute paths to other nodes. This constraint limits the approach from scaling and from distributed construction of the spanning tree. However, this constraint is released in DMSF approach since only relay node selection is needed to forward data towards collection node.

III. System Model

The nodes in vehicular networks are smart agents which interact with each other through direct ad hoc wireless links, i.e., V2V and occasionally with infrastructure or Road Side Unit (RSU), i.e., V2I. Mobile nodes mutually exchange status information which are useful for efficient mobility and traffic cooperative awareness. The focus of this paper is on the collection of data measured by the vehicles within an urban area. Therefore, we consider a geographical area with multiple roads and intersections. In the center of the area, an RSU is placed to regularly collect the data from N connected vehicles driving around. Every vehicle aims to periodically transfer a data of size M bytes to the RSU at a transmission power level P_{tr}. The obstacles and signal attenuation do not always allow direct communication between all the nodes and the RSU. Hence, a multi-hop data transfer is required.

To study the network behavior under different conditions and evaluate the ITS applications, several research studies investigated various channel propagation models of DSRC [12] [3]. In [3], authors carried out extensive measurement campaigns in different communication scenarios: the Line-of-Sight (LoS) and Non-LoS (NLoS) cases and for urban, suburban, highway, and rural areas characterized by different traffic densities. The resulting model is adopted for this paper, it is well-defined, and its parameters are explicitly given. By adopting the log-distance power law, the path loss expression is given by:

$$PL(d) = PL_0 + 10\nu \, log_{10} d + S, \quad d_{\min} \leq d \leq d_{\max}, \quad (1)$$

where d is the distance between the transmitter and the receiver in meter and ν is the path loss exponent related to the propagation environment. The parameter S models a zero-mean random variable with normal distribution and standard deviation σ_S modeling the large-scale fading. The term PL_0 is given by:

$$PL_0 = PL_0(d_0) - 10\nu \; log_{10}d_0, \tag{2}$$

where d_0 is the reference distance and $PL_0(d_0)$ is the path loss value at the reference distance. Its value is given along with other parameters in Table I. This model is derived from measurement campaigns where $d_{\min}$ and $d_{\max}$ are bounding the model validity domain.

The communication link between the UAV and the ground nodes is modeled differently. The two nodes are in LoS with a certain probability P_r which depends on the UAV's altitude h_u [4]. P_r is given by:

$$P_r = \frac{1}{1 + Bexp(-C[\theta(h_u, d_{uv}) - B])}, \tag{3}$$

where B and C are environment-dependent constants, θ represents the elevation angle and is given by $\theta = \frac{180}{\pi} sin^{-1}(\frac{h_u}{d_{uv}})$, and d_{uv} denotes the Euclidean distance between the UAV and the ground node. The probability of having a NLoS link is equal to $1 - Pr$. Hence, the path loss expression in dB of the Air-to-Ground (A2G) link is given by [4]:

$$PL^{A2G}[dB] = P_r PL^{LoS} + (1 - P_r)PL^{NLoS}, \tag{4}$$

where PL^{LoS} is the path loss effect of LoS link and is expressed in dB as $PL^{LoS} = 10\gamma log_{10}(\frac{4\pi f d_{uv}}{C_l}) + L^{LoS}$, with γ is the path loss exponent, f is the frequency of the carrier, C_l is the speed of light, and L^{LoS} is an additional attenuation of the environment. The second term of the path loss PL^{NLoS} is the effect of NLoS link and is expressed in dB as $PL^{NLoS} = 10\gamma log_{10}(\frac{4\pi f d_{uv}}{C_l}) + L^{NLoS}$ with $L^{NLoS}$is an additional attenuation for the NLoS environment.

It's important to remind here that the coherence time of the channel is very small compared to the time of data collection and construction time of the tree, therefore, we investigate the channel based on the average statistics and we consider the path loss effect only.

IV. Problem Formulation

The investigated problem can be modeled using graph theory where nodes are the graph nodes and wireless links are mapped to graph edges. At a given time, a generic undirected graph $G(V, E)$ with weighted edges results from a spatial distribution of mobile nodes. An edge matching two nodes is characterized with different attributes, namely *data rate*, *link stability*, and *closeness to sink*, which are quantified and detailed in the following subsection. The target of the next step is to leverage the link state of each node in order to compute a path from each node towards the sink in a distributed fashion. Indeed, getting information about the locations and states of all the nodes and the whole network topology in a centralized manner is an unpractical and unrealistic assumption mainly during short intervals of time. Therefore, we proceed with a distributed approach adapted to the network topology evolution. The constructed paths aim at enhancing packets delivery within a time-bounded delay while maintaining a fair exploitation of the network's resources.

A. Methodology and Metrics

The present paper focuses on the data collection task, and more specifically on a periodic data collection rounds triggered at a prefixed time period. Since every node is invited to append its own data to the received data, if available, before forwarding it towards the sink node, proceeding with broadcasting data overwhelms the network and may result in a *broadcast storm* especially in dense networks [13]. To avoid these harmful effects, every node operates through unicasting its data to a single next hop while taking into account the link states to maximize the chance of a time-bounded routing and successful delivery of packets.

To ensure a link-aware approach, three factors are considered while quantifying the weights of the wireless links. These properties are directly impacting the link quality and hence, its ability to deliver data without being distorted. Considering two nodes i and j of the network, the attributes of the link (i, j) are:

• Achieved throughput: Links with higher data rates are privileged in the construction of the data routing paths to accelerate the data collection procedure. Following is the achieved rate of link (i, j) based on Truncated Shannon Bounds (TSB) [14] and denoted by $R_{i,j}$:

$$R_{i,j} = \begin{cases} R_{max} & \text{if } SINR_{i,j} \geq SINR_{max}, \\ 0 & \text{if } SINR_{i,j} \leq SINR_{min}, \\ B\log_2(1 + SINR_{i,j}) & \text{otherwise.} \end{cases} \tag{5}$$

where R_{max} is given by:

$$R_{max} = B\log_2(1 + SINR_{max}), \tag{6}$$

where B is the system bandwidth, $SINR_{i,j}$ is the signal-to-noise-plus-interference ratio at the receiver of the link (i, j), given by:

$$SINR_{i,j} = \frac{P_{tr}H(\bar{d}(i,j))}{I + N}, \tag{7}$$

where P_{tr} denotes the transmit power level, H is the channel propagation model which is a function of average-distance between transmitter and receiver. It corresponds to the inverse of the path loss between the nodes i and j which is given in (1). I and N quantify the average interference and the noise power levels at each receiving node, respectively. Finally, $SINR_{min}$ and $SINR_{max}$ are the $SINR$ thresholds for the discretization of the data rate in DSRC technology. Their simulation values are given in Table I.

• Link stability: The relative high speed, dynamic topology, and the mobility patterns of the vehicular network impose constraints on the link stability. Hence, wireless link lifetime between two nodes is expected to persist for short time prior to be obstructed by static and/or mobile obstacles until it disappears from the available set of links. In the general case, the motion of the node in the network is hard to predict, even though some mobility patterns could be drawn in specific areas and during specific times. Hence, the position $(x_{i,t}, y_{i,t})$ of the node i at an instant t is given by:

$$(x_{i,t}, y_{i,t}) = f(x_{i,0}, y_{i,0}, t), \tag{8}$$

where f is a generic function taking as parameters the position of node i at instant $t = 0$ and the current time. Taking into account the locations, speed values, and driving directions of two given nodes at an instant t, and under the assumption that vehicles are generally driving into linear road segments with nearly constant speeds specifically for very short time-scale, the following mathematical linear system derives the expected link duration between two arbitrary vehicles i and j:

$$\begin{cases} x_{j,t} = v_{x,0}t + x_{j,0} & (9) \\ y_{j,t} = v_{y,0}t + y_{j,0} & (10) \end{cases}$$

where $x_{j,t}$ and $y_{j,t}$ are the coordinates of node j at an instant t, $x_{j,0}$ and $y_{j,0}$ are the reference coordinates of node j at $t = 0$, and $v_{x,0}$ and $v_{y,0}$ are the components of relative velocity of node j with respect to node i. With this setup, the coordinates of *node* i are given by:

$$\begin{cases} x_{i,t} = x_{i,0} & (11) \\ y_{i,t} = y_{i,0} & (12) \end{cases}$$

where $x_{i,t}$ and $y_{i,t}$ are the coordinates of node i at an instant t, $x_{i,0}$ and $y_{i,0}$ are the reference coordinates of node i at $t = 0$. From (9)–(12), the euclidean distance between the nodes enables the determination of the expected link duration, denoted by LD and returns an important indicator about the radio link stability. Indeed, the resolution of the equation below, returns the LD:

$$LD_{i,j} = t, \text{when } \sqrt{(x_{i,t} - x_{j,t})^2 + (y_{i,t} - y_{j,t})^2} = D_{\max}, \tag{13}$$

where $D_{\max}$ is the maximum communication range between the nodes i and j, which depends on the channel path loss modeled in (1).

- Closeness to sink: The main goal is to gather the data from mobile nodes driving along a geographic area to a sink. In order to reach the destination node in a reasonable time, data should progress towards the sink at every hop. Otherwise, packets may spend a long time floating from a node to another without reaching the destination. To model the progress of data towards the destination, a metric characterizing the link aptitude to approach the sink is considered. Every node is labeled with a number of hops required to reach the collection point. If a node i selects the node j as its next hop, and the latter is $\bar{N}_h(j)$ away from the destination then, the created link has an advance attribute, denoted by $adv_{i,j}$, and is given by:

$$adv_{i,j} = \bar{N}_h(j) + 1, \text{if } i \text{ selects } j \text{ as next hop.} \tag{14}$$

Once the main factors impacting the link quality are defined, the links are labeled with weights resulting from a weighted combination of the aforementioned metrics. The best subset of the set of links maximizing the overall weights is used to connect the nodes with their respective successors to reach the collection node.

B. Problem formulation

Consider that each edge of the graph $G(V, E)$, that we denote by $e_k, k = 1, \ldots, |E|$ where $|.|$ denotes the cardinality of a set, is characterized by a weight denoted by w_k. The objective is to divide the graph $G(V, E)$ into N_t disjoint trees to create a forest. This forest ensures a faster building of disjoint trees with shorter routes. The total number of trees N_t is depending essentially on the number of nodes that are within the coverage of the sink and characterized by a good link quality, e.g., $w_k \geq w_{th}$, where w_{th} is a link weight threshold defined by the operator. Define $T_t \equiv G(V_t, E_t) \subset G(V, E)$, where $t = 1, \ldots, N_t$, a tree composed of a set of vertices denoted by V_t and a set of edges denoted by E_t. Accordingly, the forest will be formed by N_t trees. Denote by $\epsilon_{k,t}$ the binary variable indicating whether an edge k of E belongs to set of edges E_t. Hence, $\epsilon_{k,t}$ is expressed as:

$$\epsilon_{k,t} = \begin{cases} 1 & \text{if } e_k \in E_t, \\ 0 & \text{otherwise} \end{cases} \tag{15}$$

Consequently, the optimization problem aiming at minimizing a multi-objective utility composed of the three metrics discussed above can be formulated as:

$$\underset{\substack{\epsilon_{k,t}, k=1,\ldots,|E| \\ t=1,\ldots,N_t}}{\text{minimize}} \sum_{k=1}^{|E|} w_k \sum_{t=1}^{N_t} \epsilon_{k,t} \tag{16}$$

subject to:

$$\epsilon_{k,t} \leq \delta_k, \tag{17}$$

$$\sum_{t=1}^{N_t} \epsilon_{k,t} \leq 1. \tag{18}$$

The solution of the optimization problem given by (16)–(18) will associate the edges e_k to the N_t trees to connect the maximum of nodes that will forward their data packets to the roots of each tree such that the total weight is minimized. Using the metrics expressed in (5), (13), and (14), the weight of a given link joining two nodes i and j and corresponding to the edge $k \in E$ can be expressed as:

$$w_k \equiv w_{i,j} = -\alpha \frac{R_{i,j}}{R_{\max}} - \beta \frac{LD_{i,j}}{LD_{\max}} + \gamma \frac{adv_{i,j}}{adv_{\max}}, \tag{19}$$

where $R_{\max}$, $LD_{\max}$, and $adv_{\max}$ are determined by Monte Carlo simulation campaigns. They are used to normalize the magnitude of the different metrics. Note that the parameter adv_{max} represents the maximum possible number of hops to reach the sink node, which is equal to the number of nodes in the network. The parameters α, β, and γ are three freedom degrees identified by the operator. Their sum is equal to one and they have to be set properly according to the operator's need. For instance, increasing α will promote the rate metric at the expense of the others. Constraint (17) ensures that an edge e_k can be considered in the association if the corresponding $SINR$ provides at least the minimum requirement for a seamless transmission. This condition is identified by the binary parameter δ_k which is given as follows:

$$\delta_k = \begin{cases} 1 & \text{if } SINR_{i,j} \geq SINR_{\min}, \\ 0 & \text{otherwise} \end{cases} \tag{20}$$

where $SINR_{i,j}$ is SINR at the receiver of the link k that can be deduced from (7) while $SINR_{\min}$ is a SINR minimum threshold. Finally, constraint (18) imposes that an edge can be associated to only one tree.

V. Data Collection with DMSF

The data collection is a specific case of more general routing problem. Indeed, for data gathering, the destination node is usually not an arbitrary node but is often the sink node having pre-defined public properties, e.g., its location is known by nodes belonging to the network. At every data collection round, the nodes have to check their link states to find the best relevant next hops in order to deliver their data content. Recall that there is no agent that has real-time global view of the network topology to execute a centralized computing of the best paths. Moreover, a given node does not have to be aware of all the network links to build its best path. Instead, it requires a local awareness of its links associating it with one-hop neighbor-nodes to make sure that its path to the sink node is loop-free. Finally, data sent by the distributed nodes should reach a subset of nodes, called the roots of the forest trees, which are responsible to relay data directly to the sink.

Algorithm V.1 Distributed Minimum Spanning Forest Algorithm

$V_R \leftarrow Best\_Sink\_Neighbors$. {The set of reliable neighbors of sink node (RSU)}
Update adjacency matrix.
$BSE \leftarrow All\_links$. {The set of basic edges}
$BRE \leftarrow \emptyset$. {The set of branch edges}
$RJE \leftarrow \emptyset$. {The set of rejected edges}
while $\| BSE \| > 0$ **do**
  $WKN \leftarrow random\_subset(V)$.
  for $n \in WKN$ **do**
    $OE \leftarrow outgoing\_edges(fragment(n))$.
    $e \leftarrow \min(OE)$.
    $BRE \leftarrow BRE \cup \{e\}$.
    {fusion fragments and update ranks}
    $frag\_1 \leftarrow right\_fragment(e)$.
    $frag\_2 \leftarrow left\_fragment(e)$.
    if $frag\_1.root \notin V_R$ & $frag\_2.root \notin V_R$ **then**
      Use default DMST update.
    else
      if $frag\_1.root \in V_R$ & $frag\_2.root \notin V_R$ **then**
        $frag\_1 \leftarrow fuse(frag\_1, frag\_2)$. {frag_1 absorbed frag_2}
      else
        if $frag\_1.root \notin V_R$ & $frag\_2.root \in V_R$ **then**
          $frag\_2 \leftarrow fuse(frag\_2, frag\_1)$. {frag_2 absorbed frag_1}
        else
          continue. {frag_2 and frag_1 could not be fused}
        end if
      end if
    end if
    $BRE \leftarrow BRE \cup \{e\}$.
    Update RJE.
    $BSE \leftarrow BSE \setminus (\{e\} \cup REJ)$.
  end for
end while

Figure 2. DMSF algorithm.

The main idea of the proposed approach is to build multiple trees with minimum overall weights and rooted at a set of predefined nodes to create a Forest of N_t trees. We call this subset of nodes, i.e., the roots, as V_R where $|V_R| = N_t$. The tree-based topology of the collection algorithm focuses on ensuring a convergence of the collected data towards the destination and using loop-free paths. Moreover, each node can join a single tree and forward its own data and other received data exclusively to one *parent* node. Working with minimum spanning forest instead of a single minimum spanning tree is preferred because it leads to short paths to reach the destination. In addition, considering multiple roots allows the network to avoid bottlenecks in the last hops and ensures a load balance throughout the network.

Figure 2 illustrates a high-level description of the proposed DMSF algorithm. For a periodic data collection, where all the vehicles in a given geofence area are expected to transmit their generated and occasionally received data, nodes should first build routing paths especially when the exchanged data size is important. In order to build loop-free paths leading to the sink node, the algorithm starts with the selection of the members of V_R. Then, nodes initiate the construction of fragments based on the best available link weights and fuse fragments into bigger ones. The process continues in a fully distributed way until all the nodes, reachable from at least one of the V_R members, join a tree of the forest. It may happen that some nodes do not join any of the forest's trees if the network graph is physically disconnected. Hence, these nodes will not transmit data until joining routing trees to save communication bandwidth.

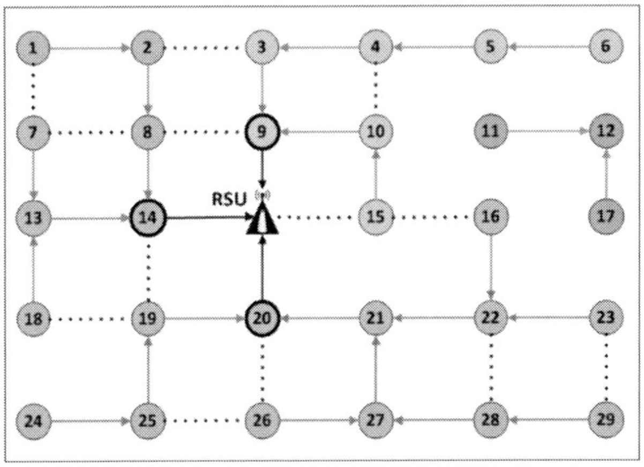

Figure 3. Illustration of routing paths resulting from DMSF algorithm for a regular vehicles distribution. The set of nodes $V_R = \{9, 14, 20\}$ represents the roots of the formed N_t trees.

Figure 3 shows an illustration of the paths built for a network graph with some disconnected parts, e.g., the tree $\{11, 12, 17\}$. The built trees reflect the fully distributed and parallel behavior of the algorithm. The root selection and rank updates of the fragments are similar to DMST algorithm. However, some updates are carried out to promote the data progress towards the RSU. These updates force the nodes member of V_R to be the roots of the formed trees. It is also worthy to mention that DMSF approach does not ensure the shortest path in terms of hop count however, it fosters the stable routes with higher throughput along with the convergence to the destination. For instance, the node 26 selects a more stable parent which is node 27 with hop-count equal to four instead of node 20 at two hops to RSU. The same scenario occurs with node 15, it chooses to go through longer route and did not join V_R because of the weak connectivity it has with the sink node.

Once the nodes finish building their routes, a data collection is initiated by the nodes joining the trees. Every node appends its locally generated data with the received data from children nodes in the routing trees, if any, and forwards data to its parent node. The roots of the trees deliver the received data to the collection point as soon as they receive data from their respective children nodes. When the next session is due, the DMSF algorithm is initiated again and the data gathering is performed by its end, and so forth. Or, in order to increase the data collection speed, new DMSFs can be created based on previously constructed DMSFs during previous session. This approach is left for a future extension of this work.

VI. Performance Analysis

This section describes the simulation environment and discusses the performance results of the proposed algorithms.

A. Simulation Setup

An urban traffic area is considered under the form of a grid network with controlled intersections. An RSU is deployed in the central junction of the studied area. The UAVs are hovering at a fixed altitude h_u above the junctions. The mobility traces of ground nodes were generated using the Simulator of Urban MObility (SUMO) [15] and the algorithms were implemented and tested in Matlab. Table I lists the simulation parameters that have been used.

TABLE I. SIMULATION PARAMETERS.

Parameter	*Value/Description*
area dimensions	1km x 1km
Number of junctions	5 x 5
Road segment length	250 m
System bandwidth	10 Mhz
Transmission power (P_{tr})	23.8 dBm
$PL_0(d_0)$	58.81 dB
Path Loss Exponent (ν)	1.83
σ_S	4.48 dB
$SINR_{min}$	-6.37 dB
$SINR_{max}$	7.35 dB
h_u	100 m

B. Performance results

In this section, the performance of the algorithms are quantified. Indeed, the data collection ratio, the required time to achieve the collection task, and the average number of hops to deliver data packets are studied for DMSF, DMST, and Greedy Perimeter Coordinator Routing (GPCR) [7] protocols. The latter is a position-based routing protocol. It leverages the planar property of the graph formed by road segments and junctions. GPCR uses the restricted greedy forwarding and a repair strategy to deal with local maximum problem [16]. Afterwards, the impact of using UAVs in assisting data collection using DMSF algorithm is evaluated for a low-density network.

1) Data Collection ratio: The considered data collection scenario in this paper is periodic and asynchronous. This means data could be concentrated in intermediate nodes which need to forward received data from children nodes to respective parent nodes. If the link quality is not considered in the selection of the next hop, many packets risk to be lost. Figure 4 shows the variation of the data collection ratio with the number of vehicles for the three algorithms. The proposed DMSF algorithm achieved clearly higher data delivery, slightly better than DMST particularly in low and medium traffic. For dense network, DMST performance drops significantly because of the increasing routing tree and the dynamic nature of the

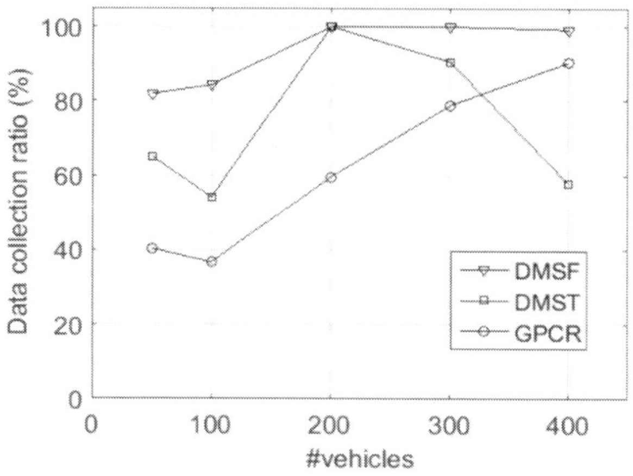

Figure 4. Data collection ratio for different traffic densities

network. GPCR is a position-based routing algorithm which forwards data to the nodes ensuring always a better progress towards the destination node. It uses a restricted greedy forwarding. When the restricted greedy forwarding fails to find a next hop, a recovery strategy such as the *right hand rule* is used. It is clear that the delivery ratio of this algorithm is limited for low-density network. However, it grows up with the network density because of the higher probability to get better relay node to reach the sink node. Even though, GPCR algorithm did not achieve DMSF packet delivery ratio.

2) Overall Delay: One major factor of data collection in mobile networks is the time bound within which the collection task has been achieved.

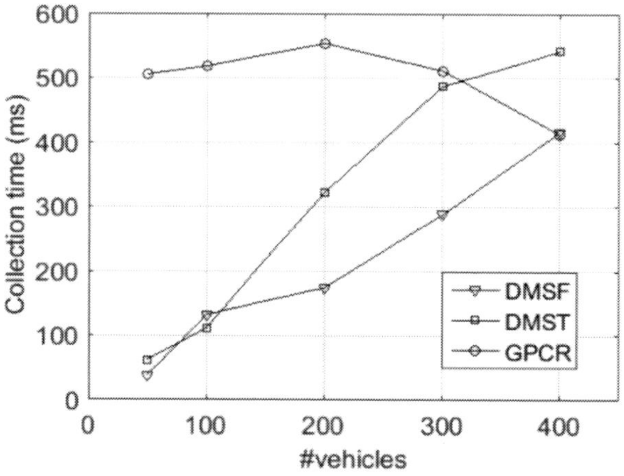

Figure 5. Data collection time for different traffic densities

The faster the collection is performed, the better the approach is. Indeed, in topology-based approaches, the computed paths are subject to change within a short time. Hence, a slow data forwarding will probably face outdated paths and end up with discarding packets or recomputing the path. Figure 5 shows the collection time in milliseconds to route the data from all the network nodes towards the sink. Of course, only nodes able to find a relay will send data. The other nodes will discard their data and wait to the next data collection session. For different densities of the network, the DMSF performs better than DMST, which is, in its turn, performing better than GPCR in terms of overall data collection delay for low and medium traffic densities. In dense network, GPCR overall delay outperforms DMST. In GPCR, the extra delay is caused by the long paths traveled by some packets when entering the recovery mode. The difference between DMSF and DMST in terms of collection time is due to the partitioned routing structures (i.e., trees) in DMSF leading to load balance and bottleneck avoidance in the neighborhood of the sink node. These factors lower the packet processing time in intermediate nodes, and hence lower the end-to-end journey delay of the packets.

3) Average number of hops: The average number of hops denotes the length of the paths of delivered packets traveled through. Figure 6 shows that GPCR presents shortest paths in average. This is due to low delivery ratio when the network is sparse, and the greedy nature of the protocol to approach the sink node in dense network. For DMSF, the results are slightly higher because of the consideration of the link quality and the distributed nature of the paths construction. DMSF performs better than DMST because the latter uses a single routing tree causing the leaf nodes to be more distant from the tree root.

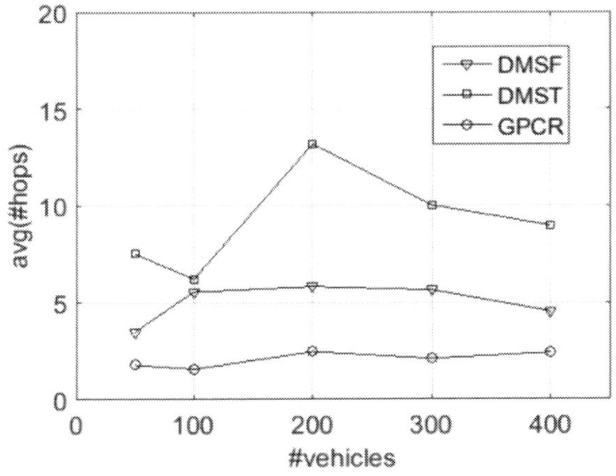

Figure 6. Average hops' number for different traffic densities

4) UAVs impact: For low density traffic, the network is fragmented. This causes a degraded performance of the data collection. Exploiting the existing UAVs as relay nodes between two fragments of the network has a positive impact on the system performance. Figures 7-9 depict the impact of using UAVs with DMSF algorithm in low-density network compared to DMSF with only ground nodes. The hovering of UAVs above the junctions allow them to link the network fragments and hence increase the packet delivery ratio which approaches 98% when using 24 drones. The overall delay and the average number of hops within the network did not change significantly which means the the UAVs enhance the network behavior with a very low cost.

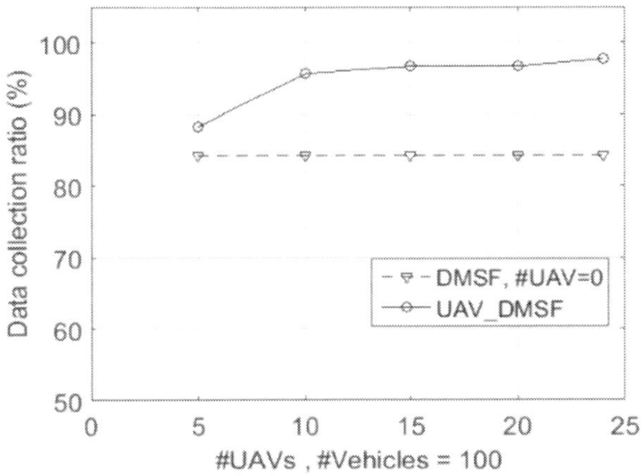

Figure 7. Impact of UAVs on data collection ratio

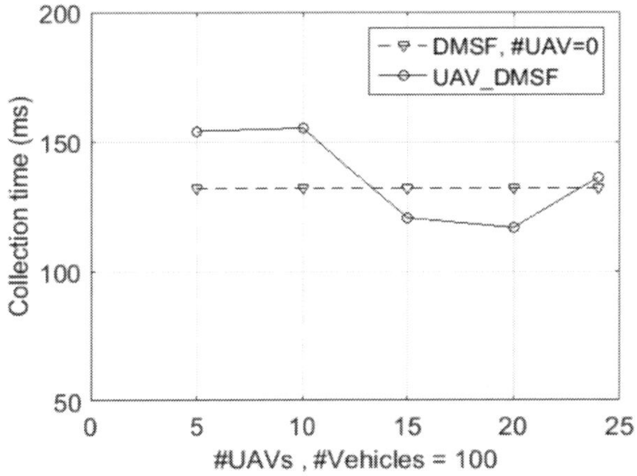

Figure 8. Impact of UAVs on data collection time

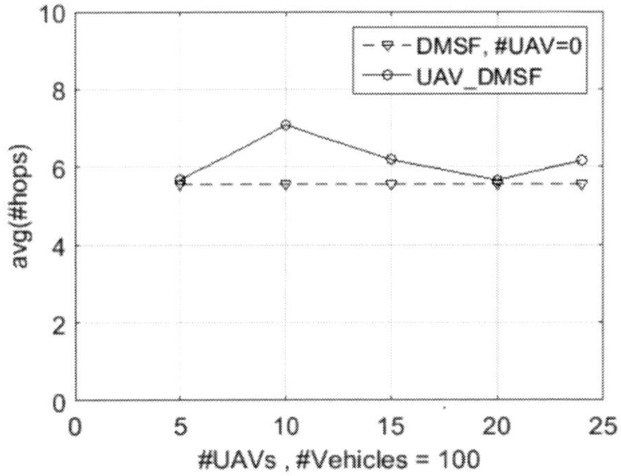

Figure 9. Impact of UAVs on average hops' number

VII. Conclusion

Vehicular network enables a set of applications which make the driving experience safer and more efficient. Mobile nodes continuously transmit data which is gathered by remote devices. In this paper, the data collection challenge had been tackled in an urban area through the proposal of DMSF algorithm. The proposed solution achieves a very high data delivery ratio and scales better than GPCR and DMST algorithms in terms of collection time. The provided assistance of UAVs in relaying data between disconnected network fragments clearly enhances the data collection task with a very low cost in terms of overall delay. In a broader view, the present solution could help in enhancing the network nodes connectivity and end-to-end data delivery by assuring a local link-aware next-hop selection.

Acknowledgment

This publication was made possible by NPRP grant #NPRP8-2459-1-482 from the Qatar National Research Fund (a member of Qatar Foundation). The statements made herein are solely the responsibility of the authors.

References

[1] "ETSI TS 101 539-3 V1.1.1(2013-11) - Intelligent Transport Systems (ITS); V2X Applications; Part 3: Longitudinal Collision Risk Warning (LCRW)application requirements specification," http://www.etsi.org/deliver/etsi_ts/101500_101599/10153903/01.01.01_60/ts_10153903v010101p.pdf, NOV 2013, accessed: 2018-04-21.

[2] "ETSI TS 101 539-3 V1.1.1(2013-11) - Intelligent Transport Systems (ITS); V2X Applications; Part 1: Road Hazard Signalling (RHS) application requirements specification," http://www.etsi.org/deliver/etsi_ts/101500_101599/10153901/01.01.01_60/ts_10153901v010101p.pdf, AUG 2013, accessed: 2018-04-21.

[3] H. Fernández, L. Rubio, and V. M. Rodrigo-Peñarrocha, "Path loss characterization for vehicular communications at 700 MHz and 5.9 GHz under LOS and NLOS conditions," IEEE Antennas and Wireless Propagation Letters, vol. 13, May 2014, pp. 931–934, DOI: 10.1109/LAWP.2014.2322261.

[4] H. Ghazzai, A. Feidi, H. Menouar, and M. L. Ammari, "An exploratory search strategy for data routing in flying ad hoc networks," in Proceedings of the IEEE 28th Annual International Symposium on Personal, Indoor, and Mobile Radio Communications (PIMRC) Oct 8–13, 2017, Montreal, QC, Canada, 2017, pp. 1–7, DOI: 10.1109/PIMRC.2017.8292474.

[5] H. Ghazzai, H. Menouar, and A. Kadri, "Data Routing Challenges in UAV-assisted Vehicular Ad hoc Networks," in Proceedings of VEHICULAR 2017: the 6th International Conference on Advances in Vehicular Systems, Technologies and Applications July 23–27, 2017, Nice, France, 2017, pp. 85–90, ISBN: 978-1-61208-573-9.

[6] R. G. Gallager, P. A. Humblet, and P. M. Spira, "A Distributed Algorithm for Minimum-Weight Spanning Trees," Journal ACM Transactions on Programming Languages and Systems (TOPLAS), vol. 5, no. 1, Jan. 1983, pp. 66–77, DOI: 10.1145/357195.357200.

[7] J. Liu et al., "A survey on position-based routing for vehicular ad hoc networks," Springer Telecommunication Systems, vol. 62, no. 1, May 2016, pp. 15–30, Online ISSN: 1572-9451.

[8] S. Ucar, S. C. Ergen, and O. Ozkasap, "Multihop-Cluster-Based IEEE 802.11p and LTE Hybrid Architecture for VANET Safety Message Dissemination," IEEE Transactions on Vehicular Technology, vol. 65, no. 4, Apr. 2016, pp. 2621–2636, DOI: 10.1109/TVT.2015.2421277.

[9] H. Huang, L. Libman, and G. Geersp, "An Agent Based Data Collection Scheme for Vehicular Sensor Networks," in Proceedings of the IEEE 24th International Conference on Computer Communication and Networks (ICCCN) August 3–6, 2015, Las Vegas, NV, USA, 2015, DOI: 10.1109/ICCCN.2015.7288374.

[10] A. Soua and H. Afifi, "Adaptive data collection protocol using reinforcement learning for VANETs," in Proceedings of the IEEE 9th International Wireless Communications and Mobile Computing Conference (IWCMC) July 1–5, 2013, Sardinia, Italy, 2013, pp. 1040–1045, DOI: 10.1109/IWCMC.2013.6583700.

[11] P. Fazio, F. D. Rango, C. Sottile, and A. F. Santamaria, "Routing Optimization in Vehicular Networks: A New Approach Based on Multiobjective Metrics and Minimum Spanning Tree," International Journal of Distributed Sensor Networks, vol. 9, no. 11, 2013, pp. 1–13, DOI: 10.1155/2013/598675.

[12] A. Al-Hourani, S. Chandrasekharan, G. Baldini, and S. Kandeepan, "Propagation measurements in 5.8GHz and pathloss study for CEN-DSRC," in Proceedings of the IEEE International Conference on Connected Vehicles and Expo (ICCVE) Nov 3–7, 2014, Vienna, Austria, 2014, pp. 1086–1091, DOI: 10.1109/ICCVE.2014.7297518.

[13] N. Wisitpongphan, O. K. Tonguz, and J. Parikh, "Broadcast storm mitigation techniques in vehicular ad hoc networks," IEEE Wireless Communications, vol. 14, no. 6, Dec. 2007, pp. 84–94, DOI: 10.1109/MWC.2007.4407231.

[14] A. Burr, A. Papadogiannis, and T. Jiang, "MIMO Truncated Shannon Bound for System Level Capacity Evaluation of Wireless Networks," in Proceedings of the IEEE Wireless Communications and Networking Conference Workshops (WCNCW) July 1–1, 2012, Paris, France, 2012, pp. 1040–1045, DOI: 10.1109/WCNCW.2012.6215504.

[15] D. Krajzewicz, J. Erdmann, M. Behrisch, and L. Bieker, "Recent development and applications of SUMO - Simulation of Urban MObility," International Journal On Advances in Systems and Measurements, vol. 5, no. 3&4, Dec. 2012, pp. 128–138.

[16] C. Lochert et al., "Geographic Routing in City Scenarios," ACM SIGMOBILE Mobile Computing and Communications Review, vol. 9, no. 1, Jan. 2005, pp. 69–72, DOI: 10.1145/1055959.1055970.

LiDAR-based SLAM algorithm for indoor scenarios

Felipe Jiménez, Miguel Clavijo and Javier Juana
University Institute for Automobile Research (INSIA)
Universidad Politécnica de Madrid (UPM)
Madrid, Spain
e-mails: {felipe.jimenez, miguel.clavijo}@upm.es, javier.juana.serrano@alumnos.upm.es

Abstract— **Simultaneous Localization and Mapping (SLAM) algorithms are one of the elements that have great relevance for autonomous driving in order to locate the vehicle, even in areas in which other methods have difficulties, or to improve the positioning given by these other systems. It also offers knowledge of the scenario in which the vehicle moves, information that can have multiple uses. There are several solutions to the SLAM problem using Light Detection and Ranging (LiDAR), but these algorithms require a high computational cost. However, certain environments with a specific structure allow the use of more simplified algorithms. Specifically, this paper shows a SLAM algorithm where only the LiDAR signal is used and vertical planes are taken as reference (perpendicular to the ground plane). This solution is quite effective in some scenarios, such as indoor parking areas. In addition, various alternatives are explored to increase the robustness of the results of positioning and mapping reconstruction. The algorithm has been tested in real scenarios with satisfactory results.**

Keywords - SLAM; LiDAR; Autonomous vehicle; Detection; Algorithm.

I. INTRODUCTION

One of the biggest difficulties of autonomous vehicles is the ability to react correctly and safely to the eventualities that may occur during driving. Consequently, these vehicles must be equipped with sensors capable of responding in a very short time and a computer capable of interpreting all this information in real time. Among these sensors are cameras, radars, Global Positioning System (GPS), LiDAR sensors, infrared sensors, etc. Specifically, LiDAR has advantages that make it particularly suitable for autonomous driving [1]. One of the main advantages of LiDAR sensors versus computer vision is that the latter are very sensitive to light changes or environmental conditions [2]. In addition, cameras are directional and only detect objects in the direction in which they are placed, and LiDAR sensors offer most of the time a 360° field of view, obtaining a complete point cloud from the vehicle's surroundings. On the other hand, a drawback of the LiDAR sensors is that the density of points decreases with distance due to the divergence between the laser beams.

Solving the SLAM problem consists in the estimation of the movement of the ego-vehicle and the mapping of the environment in which it is located simultaneously. Among the possible applications of this technique, we can highlight its use for autonomous vehicle guidance systems and the creation of three-dimensional models of the environment in a fast and accurate manner. The growing interest in self-driving cars has caused researchers to once again seek a solution to the problem of SLAM, a problem that appeared in robotics more than 25 years ago. This concern to achieve an accurate SLAM solution is fundamentally due to the need for an error positioning of centimeters and an understanding of the environment for a proper path-planning and decision making. As far as the SLAM techniques are concerned, there are many branches to tackle this problem. From the classical techniques used in robotics, or new approaches using range sensors, or techniques based on artificial intelligence, to new trends, such as multivehicle SLAM [3].

When talking about the problem of vehicle localization, the solution is not trivial using a GPS. GPS has shown great performance for the location, however, cannot guarantee an error below centimeters in all cases, even when using D-GPS with a positioned base station. A classic approach would be to make use of inertial units, encoders on the wheels, etc. However, all of them produce an incremental error in time, which makes a correct localization impossible. Another classic approach could be the localization of the ego-vehicle using the line markings on the road, but these are not enough when the vehicle approaches complex scenarios, like a crossing or merging lanes. Finally, we always look for a trade-off, merging different approaches into a fusion.

Other approaches would be techniques based on Visual-SLAM or LiDAR-based SLAM, both make use of the extraction of characteristic points of the environment, or landmarks. After the extraction of these landmarks, a matching and calculation of the relative movement are carried out. The prediction of the new position of the vehicle can be estimated, on one hand, by filter techniques, e.g., Extended Kalman Filter (EKF) [4]–[6] or Particle Filter [7]–[9]. On the other hand, we could also find another approach for the same task, this time using optimization methods, such as Bundle Adjustment [10] among others.

More recently, the use of artificial intelligence techniques, specifically deep learning, has gained great interest due to the great advance of computing capacity and the development of several quality databases (e.g., KITTI dataset [11], Ford Campus [12] or Málaga Urban data [13]). The use of Convolutional Neural Networks (CNNs) is very suitable for the recognition of images and for the extraction of characteristics. On the other hand, CNNs have not only been used to extract image characteristics but also to estimate continuous signals when dealing with regression problems. Therefore, we can find several developments where they take advantage of these benefits to apply to the SLAM problem [14]–[17].

Although the SLAM problem is not new and there are many developments in the field, it still presents some issues. The first one is the drift that occurs when the trajectory increases, and on the other hand, the construction of the map in any weather condition, traffic or time of year [18].

In indoor scenarios, the GPS signal is not available or is not sufficiently reliable and its accuracy is low, so it cannot be used as the only reference for the positioning of autonomous vehicles. Autonomous vehicles are equipped with perception systems, being the LiDAR one of the best ones regarding robustness and information provided. This sensor has been used in common SLAM problems, although these algorithms require a high computational cost. However, certain environments with a specific structure allow the use of more simplified algorithms. Specifically, this paper shows a SLAM algorithm where only the LiDAR signal is used, and vertical planes are taken as reference (perpendicular to the ground plane). In addition, various alternatives are explored to increase the robustness in the result of positioning and scenario reconstruction.

The rest of the paper is structured as follows. In Section 2, the SLAM algorithm developed in this paper is described. In Section 3, results from several tests both indoors and outdoors are depicted. Finally, in Section 4, conclusions and future works are presented.

II. SLAM ALGORITHM

The proposed algorithm for scenarios where the vertical planes can be taken as reference includes 3 phases:

- Detection of the characteristic elements of the environment (vertical planes)
- Determination of the trajectory (the planes detected are projected on the horizontal plane and the lines resulting from those projections and their intersections are used)
- Reconstruction of the environment (three-dimensional reconstruction by superimposing point clouds)

A. Characteristic elements detection

The characteristic elements are those elements that have certain properties that allow them to differentiate themselves from other elements of the environment. Therefore, the repetitiveness with which they are detected will be very important to ensure that each of them is detected in successive time intervals.

In this way, they can be tracked over time and the trajectory of the vehicle can be determined from the relative movement of the detected characteristic elements.

Therefore, the detection of these elements is the starting point of the algorithm, and the precision in their detection determines the accuracy of the trajectory obtained.

In the urban and industrial environment, for which this algorithm is designed, walls and columns of rectangular section can be found abundantly. Therefore, the characteristic elements to be detected are vertical corners and vertical planes.

However, due to the difficulty of detecting the vertical corners directly, only the vertical planes will be detected, and the corners will be extracted from them.

Method 1: Lines detection in each laser layer

For the detection of characteristic elements, a coefficient has been implemented, proposed in [19], which evaluates the smoothness of a surface, with the aim of detecting vertical corners. However, this coefficient has finally been discarded because all those rough objects that were detected in the environment could be detected as possible landmarks.

As an alternative, for the detection of sudden changes of curvature in the different sections, the angle formed by the segment between two consecutive points with the horizontal axis of the XY plane has been analyzed.

A more accurate way to detect these lines is to calculate a local line for each point. Then, those points that are placed a certain distance away from the line are removed, and the equation of the line is obtained with the others and the average square error is recalculated. If the error is less than a certain threshold value, the line is accepted. This process is repeated for all laser layers. Once all the lines have been extracted, they are compared with each other to detect possible matches between one layer and the others.

Method 2: Planes detection from points clouds

Another method involves extracting the planes in a direct way from the point cloud in 3D. For this purpose, the M-estimator SAmple Consensus (MSAC) algorithm has been used [20]. The MSAC algorithm is a variant of the RANdom SAmple Consensus (RANSAC) algorithm. Once the model that we want to fit is known, in this case a plane, this algorithm optimizes according to the number of inliers and outliers with the cost function (1):

$$Cost = \sum_i \rho\,(e_i^2) \tag{1}$$

$$\rho(e^2) = \begin{cases} e^2 & e^2 < T^2 \\ T^2 & e^2 \geq T^2 \end{cases} \tag{2}$$

Where T is the threshold for considering inliers and e^2 provides the error for the point data.

The adjustment is executed and the points of the extracted plane are eliminated for the next execution. From the set of all the extracted planes, only those perpendicular to the horizontal plane of the vehicle are of interest for the calculation of the trajectory. Therefore, planes that do not meet this condition or those whose average error is too high are removed.

B. Trajectory calculation

The landmarks that are used to calculate the trajectory are the vertical planes of the environment. These planes are projected on the horizontal plane and, in order to determine the trajectory we work with the lines resulting from that projection.

In addition to the lines that result from the projection of the vertical planes, the points of the intersections of these lines are also used for the calculation of the trajectory. Some of these points correspond to the real corners of the planes that share a corner in the field of vision of the LiDAR, while the rest correspond to the virtual corners that result from the prolongation of those planes.

The mathematical representation of intersecting lines is presented with 2 notations based on their orientation:

$$\begin{cases} \text{Type 0:} & y = m_0 x + n_0 \\ \text{Type 1:} & x = m_1 y + n_1 \end{cases} \tag{3}$$

Once the straight lines have been calculated, for the calculation of the intersections (Figure 1), the four possible pairings have been considered according to the type of line. The number of intersections for a number n of lines is determined by the following expression:

$$\mathrm{n_{intersections}} = \frac{n}{2} \cdot (n-1) \tag{4}$$

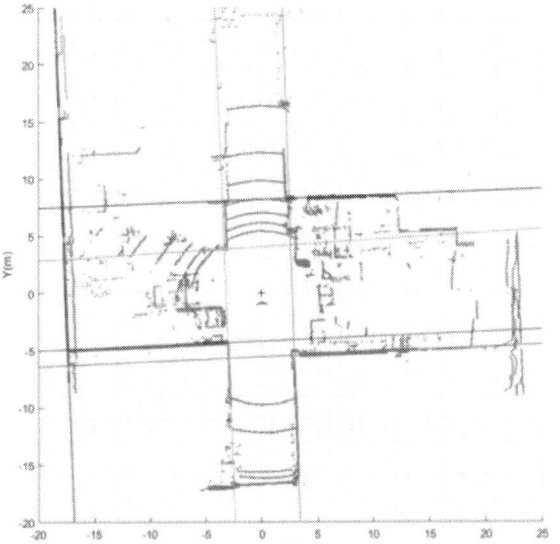

Figure 1. Lines intersections

Intersections located further than 100 m are eliminated.

To determine the trajectory of the vehicle, three different methods have been developed to obtain the values of displacement and rotation.

Method A:

The first one uses a function called *estpos* which calculate the displacement and the rotation of the vehicle simultaneously, from the relative movement of the intersections previously calculated, by solving a linear system of equations.

First, those points that have been detected in two consecutive frames are identified and matched. Therefore, the points of the previous instant, called *points0*, are displaced and rotated with the values of dx_0, dy_0 and φ_0 calculated at the previous instant. Then, the nearest point of the current frame is searched for each of the points already displaced. If the distance is less than a threshold, both points are considered to correspond to the same characteristic element and the pair of points is saved.

Once the points detected in the two frames have been determined, the longitudinal and lateral displacements and the yaw angle are calculated. For this, the following system of equations is solved.

$$\begin{bmatrix} \cos\varphi & -\sin\varphi \\ \sin\varphi & \cos\varphi \end{bmatrix} \cdot \begin{bmatrix} x_0 \\ y_0 \end{bmatrix} + \begin{bmatrix} dx \\ dy \end{bmatrix} = \begin{bmatrix} x_1 \\ y_1 \end{bmatrix} \tag{5}$$

Where x_1 and y_1 correspond to the coordinates of the points in the current frame, and x_0 and y_0 correspond to the coordinates of the points in the previous frame. The unknowns of this system are: φ, dx and dy.

This system is a non-linear system and for its resolution, it would be necessary to resort to iterative calculation methods. However, since the time interval between two consecutive frames is very small, it can be considered that the angle φ will also take small values.

In this way, simplifications can be considered and the system becomes linear. The simplified system is as follows.

$$\begin{bmatrix} 1 & 0 & -y_0 \\ 0 & 1 & x_0 \end{bmatrix} \cdot \begin{bmatrix} dx \\ dy \\ \varphi \end{bmatrix} = \begin{bmatrix} x_1 - x_0 \\ y_1 - y_0 \end{bmatrix} \tag{6}$$

Method B:

The second method is a variation of the previous one, and instead of solving the linear system of equations, it uses the Iterative Closest Point (ICP) algorithm to obtain the displacement and rotation values. The pairing of the points occurs identically.

Method C:

Finally, the third method calculates the rotation and displacement of the vehicle independently. On the one hand, the angle rotated is calculated using the equations of the straight lines and later, the displacement of the vehicle is calculated from the intersections.

To do this, the first step is to match each line with its line equivalent to the previous frame. This process is identical to the pairing of points, with the difference of using the equation of the line instead of the coordinates of the point.

C. Scenario reconstruction

In this step, it is necessary to transfer and rotate the point cloud obtained in each of the frames, and in this way generate a single three-dimensional image that includes all the points in absolute coordinates.

This reconstruction is of special interest because it is used to quantify the precision of the trajectory calculation and, in this way, to be able to compare the different algorithms used and the influence of the improvements introduced in them.

D. Algorithm improvements

On the previous algorithm, 2 improvements have been proposed:

- Weighting of points and lines.
- Correction of the possible inclination of the laser with respect to the ground.

The first correction is based on the fact that not all data is extracted with the same reliability. Therefore, a confidence value is assigned to each of the points or the lines, depending on certain characteristics that may affect the accuracy with which they have been calculated.

This weighting is carried out by increasing the number of points based on a confidence value. In this way, those points with a higher confidence coefficient will be repeated a greater number of times than those with a lower coefficient.

For the definition of the value of the confidence of the points, two factors have been taken into account: the distance of the points from the vehicle and the angle formed by the lines with which the point has been calculated.

The distance of the points has been quantified by the inverse of the distance. Since x_p and y_p are the coordinates of the points, the weighting coefficient has been defined as follows:

$$C_1 = \frac{1}{\sqrt{{x_p}^2 + {y_p}^2}} \quad (7)$$

In this way, the points near the vehicle will have a greater weight compared to those further away.

On the other hand, because the points have been obtained by intersecting the lines corresponding to the planes detected, the greater the angle formed by these lines, the greater the precision with which the point has been calculated. To quantify this effect, the angle formed by these lines has been calculated and the following coefficient has been used.

$$C_2 = 1 - \frac{2 \cdot \left(\frac{\pi}{2} - \alpha\right)}{\pi} \quad (8)$$

where α is the angle formed by the lines with which each point has been calculated.

Finally, a third coefficient that combines the previous two has been implemented.

$$C_3 = C_1 \cdot C_2 \quad (9)$$

On the other hand, the lines resulting from the projection of the detected planes are weighted taking into account two factors: the number of points contained in the plane and the error in obtaining it.

In this way, a confidence value has been defined that is a function of the value of the mean square error when obtaining the plane and the number of points of the same.

$$C = n_{points} \cdot (1 - \varepsilon) \quad (10)$$

Where n_{points} are the points of each of the planes and ε is the mean square error calculated by the pcfitplane function.

Regarding the second improvement, it is intended to correct the lack of parallelism between the plane of the laser and the ground. Generally, in garages, placing the laser level on the roof of the car, the planes detected are perpendicular to the horizontal plane of the laser. However, this inclination is not negligible. To avoid this effect, a plane is defined that remains constant regardless of the possible rolling and pitching movements of the vehicle, or the possible inclination of the terrain. Therefore, a reference plane is defined as perpendicular to the detected vertical planes.

To sum up, the flowchart of the implemented algorithm is depicted in Figure 2.

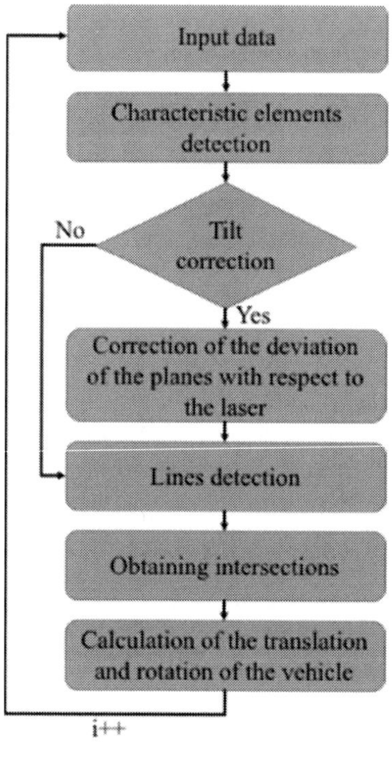

Figure 2. Algorithm flowchart

III. TESTS AND RESULTS

Different tests have been carried out not only in outdoors scenarios, but also in indoor scenarios using an instrumented vehicle that includes a 3D laser scanner Velodyne VLP-16. The LiDAR sensor was placed on top of the car in order to acquire a 360º field of view.

The tests have been carried out in 3 different scenarios:

- INSIA laboratory (Figure 3a). It is a particularly suitable environment for the calculation of the trajectory since it is an environment with a large number of vertical walls and some columns. In this scenario, 2 types of maneuvers have been carried out: straight manoeuvre and L-shape manoeuvre.

Figure 3. a) INSIA laboratory, b) Parking area, c) Narrow urban street

- Indoor parking area (Figure 3b). This type of environment also has columns and some vertical walls. However, there is not much quantity of them and being an environment with a low ceiling, a lot of points are lost in it. In addition, parked cars are also an obstacle to detection.
- Narrow urban street (Figure 3c). this method will work correctly in those streets where there are a certain number of intersections.

A. Final method selection

The first of the scenarios has been used to determine the most appropriate method among those presented for the calculation of the character elements, the determination of the turn and the translation, as well as in the alternatives of improvements over the algorithm. To do this, the bag is taken with L-shape trajectory and 7 points have been chosen from the main walls of the ship in which the maximum distance between the same wall detected in different frames is measured. The chosen points are those marked with a yellow square in Figure 4.

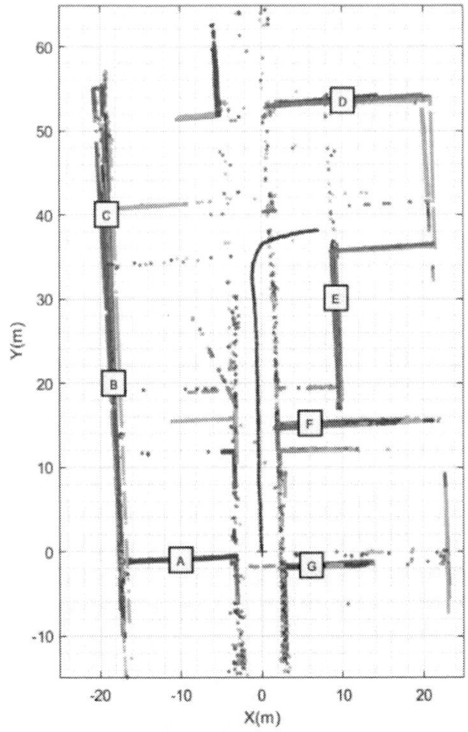

Figure 4. Control points for error calculation

First, we analyze the 2 methods to obtain the landmarks. It is observed as the first option of calculation of lines by layers has limitations in the detection of the planes from a certain distance. However, the function based on the plane-fitting is able to identify them and with a much lower computational time. Therefore, Method 2 is chosen from those described in section 2A.

To calculate the displacement and rotation of the vehicle, the three methods described above were tested, verifying that only method C provides null errors in ideal environments and, therefore, the best results in real environments.

Finally, the improvements proposed for the algorithm have been analyzed. The results are shown in Table I.

TABLE I. COMPARISON OF IMPROVEMENTS ALTERNATIVES

Improved method	Average error	Relative error
Base case (without improvements)	0,76 m	1,65 %
Points weighing. Distance	0,61 m	1,32 %
Points weighing. Angle	0,60 m	1,30 %
Points weighing. Combined	0,59 m	1,28 %
Lines weighing	0,86 m	1,86 %
Horizontal plane correction	0,57 m	1,24 %
Combined correction (except Lines weighing)	0,54 m	1,17 %

B. Results in real scenarios

Finally, the 3 test environments were reconstructed using the selected method. Similarly to the preliminary trials, a set of control points have been defined to evaluate the accuracy of the reconstruction. Table II shows the results of this quality indicator and Figure 5 shows the reconstructions.

TABLE II. QUALITY INDEX IN RECONSTRUCTION

Scenario	Average error	Distance	Relative error
INSIA lab (straight trajectory)	0,35 m	64, 99 m	0,54 %
INSIA lab (L-shape trajectory)	0,54 m	46,13 m	1,17 %
Parking area	0,15 m	11, 94 m	1,24 %
Urban area	0,27 m	70,91 m	0,38 %

The lowest relative errors have been obtained in the INSIA laboratory when the trajectory is straight and in the urban circulation. Both environments are characterized by having large vertical planes, corresponding to the high walls of the workshop and the facades of the buildings, respectively. In this way, these planes when detected with a great number of points, the accuracy is higher, and thanks to this the calculation of the trajectory is also more precise.

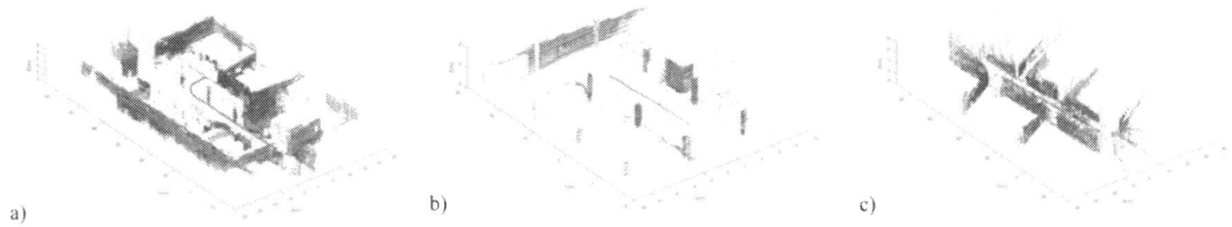

Figure 5. a) INSIA laboratory, b) Indoor parking area, c) Urban area

It is noted that, although in all cases the measurements of a Trimble BX935-INS multi-constellation GNSS receiver have been available, its signal has not been reliable due to the specific scenarios considered, so it cannot be taken as a reference.

IV. CONCLUSION

As conclusion, the developed SLAM technique requires specific characteristics of the environment in order to function correctly. Specifically, it requires the presence of a sufficient number of vertical planes, which will be used to calculate the relative movement of the vehicle. This fact happens in several real scenarios.

As an advantage, we could emphasize that it is a totally autonomous method, since it uses only the data obtained from the LiDAR, and therefore, it is not subject to the need to receive external signals as it happens to other positioning systems, such as GPS. Thanks to this, this algorithm can be used anywhere, and regardless of whether the environment is underground or outdoors, as long as it is an environment with the characteristics described above. The SLAM algorithm can be used to calculate the position of the vehicle in those areas where the GPS signal does not arrive or is weak. In addition, thanks to the simplifications introduced, its computational load is much lower than conventional SLAM algorithms, favoring its execution in real time.

The main limitation presented by the algorithm, as indicated, is the inability to calculate the displacement and rotation of the vehicle in those moments of time when the algorithm does not detect a sufficient number of characteristic elements. To solve these problems, it would be interesting to incorporate additional positioning systems such as INS, to compare the results with those obtained by the SLAM technique. Another alternative consists in incorporating other types of characteristic elements, such as, for example, the edges of the vertical planes, the poles of the traffic signs or the trunks of the trees.

Also, to facilitate detection, it could be interesting to incorporate a second LiDAR sensor. Thus, objects would be detected with a greater number of points, which would facilitate their detection.

ACKNOWLEDGMENT

This work has received the support of the MINECO CAV project (TRA2016-78886-C3-3-R), and Excellence Network SEGVAUTO-TRIES.

REFERENCES

[1] F. Jiménez, *Intelligent Road Vehicles: Enabling Technologies and Future Developments*. Elsevier, 2017.

[2] F. Jiménez and J. E. Naranjo, "Improving the obstacle detection and identification algorithms of a laserscanner-based collision avoidance system," *Transp. Res. Part C Emerg. Technol.*, vol. 19, no. 4, pp. 658–672, 2011.

[3] E. Nettleton, S. Thrun, H. Durrant-Whyte, and S. Sukkarieh, "Decentralised SLAM with low-bandwidth communication for teams of vehicles," in *Field Service Robot*, 2006, pp. 179–188.

[4] P. Newman, D. Cole, and K. Ho, "Outdoor SLAM using visual appearance and laser ranging," in *Proceedings 2006 IEEE International Conference on Robotics and Automation, 2006. ICRA 2006.*, pp. 1180–1187.

[5] A. J. Davison, "Real-time simultaneous localisation and mapping with a single camera," in *Proceedings Ninth IEEE International Conference on Computer Vision*, 2003, pp. 1403–1410 vol.2.

[6] J. M. M. Montiel, J. Civera, and A. J. Davison, "Unified Inverse Depth Parametrization for Monocular SLAM."

[7] M. Montemerlo, S. Thrun, D. Koller, and B. Wegbreit, "FastSLAM: A factored solution to the simultaneous localization and mapping problem," *Proc. 8th Natl. Conf. Artif. Intell. Conf. Innov. Appl. Artif. Intell.*, vol. 68, no. 2, pp. 593–598, 2002.

[8] M. Mohan and K. MadhavaKrishna, "Mapping large scale environments by combining Particle Filter and Information Filter," in *2010 11th International Conference on Control Automation Robotics & Vision*, 2010, pp. 1000–1005.

[9] D. Hahnel, W. Burgard, D. Fox, and S. Thrun, "An efficient fastslam algorithm for generating maps of large-scale cyclic environments from raw laser range measurements," in *Proceedings 2003 IEEE/RSJ International Conference on Intelligent Robots and Systems (IROS 2003) (Cat. No.03CH37453)*, vol. 1, pp. 206–211.

[10] Z. Zhang and Y. Shan, "Incremental motion estimation through modified bundle adjustment," in *Proceedings 2003 International Conference on Image Processing (Cat. No.03CH37429)*, vol. 3, p. II-343-6.

[11] A. Geiger, P. Lenz, C. Stiller, and R. Urtasun, "Vision meets Robotics: The KITTI Dataset."

[12] G. Pandey, J. R. McBride, and R. M. Eustice, "Ford Campus vision and lidar data set," *Int. J. Rob. Res.*, vol. 30, no. 13, pp. 1543–1552, Nov. 2011.

[13] J.-L. Blanco-Claraco, F.-Á. Moreno-Dueñas, and J. González-Jiménez, "The Málaga urban dataset: High-rate stereo and LiDAR in a realistic urban scenario," *Int. J. Rob. Res.*, vol. 33, no. 2, pp. 207–214, Feb. 2014.

[14] K. Konda and R. Memisevic, "Learning Visual Odometry with a Convolutional Network," in *International Conference on Computer Vision Theory and Applications*, 2015, pp. 486–490.

[15] M. Clavijo, F. Serradilla, J. E. Naranjo, F. Jiménez, and A. Díaz, "Application of Deep Learning to Route Odometry Estimation from LiDAR Data," in *VEHICULAR*, 2017, pp. 60–65.

[16] A. Nicolai, R. Skeele, C. Eriksen, and G. A. Hollinger, "Deep Learning for Laser Based Odometry Estimation," in *Science and Systems Conf. Workshop on Limits and Potentials of Deep Learning in Robotics*, 2016.

[17] S. Wang, R. Clark, H. Wen, and N. Trigoni, "DeepVO : Towards End-to-End Visual Odometry with Deep Recurrent Convolutional Neural Networks," pp. 2043–2050, 2017.

[18] G. Bresson, Z. Alsayed, L. Yu, and S. Glaser, "Simultaneous Localization And Mapping: A Survey of Current Trends in Autonomous Driving," *IEEE Trans. Intell. Veh.*, vol. XX, no. X, pp. 1–28, 2017.

[19] J. Zhang and S. Singh, "Low-drift and real-time lidar odometry and mapping," *Auton. Robots*, no. October 2014, 2016.

[20] P. H. S. Torr and A. Zisserman, "MLESAC: A new robust estimator with application to estimating image geometry," *Comput. Vis. Image Underst.*, vol. 78, no. 1, pp. 138–156, 2000.

Wide Transmission of Proxy Cooperative Awareness Messages

Masahiro Kitazawa, Manabu Tsukada, Hideya Ochiai and Hiroshi Esaki
Graduate School of Information Science and Technology
University of Tokyo, Japan
Email: {ktzw, tsukada, jo2lxq}@hongo.wide.ad.jp, hiroshi@wide.ad.jp

***Abstract*—Recently, autonomous vehicles have become a reality. Most of the research related to autonomous vehicles focuses on the sensors attached to vehicles. However, owing to blind spots, sensors are insufficient for avoiding accidents. Cooperative Intelligent Transport Systems (CITSs) have been introduced to reduce blind spots. These systems wirelessly communicate with other CITS-enabled vehicles and collect road and traffic information. There are two types of messages used in CITS: Cooperative Awareness Messages (CAMs) and Decentralized Environmental Notification Messages (DENMs). CAMs are used to notify the existence of sender. Some issues are known to exist with these messages. Several methods have been proposed to help overcome these issues. One of these methods is Proxy CAM. In this study, we will describe four problems of Proxy CAM and propose Grid Proxy CAM, which builds a network of Proxy CAM devices and forwards Proxy CAM packets. We have been developing two methods for this system. The first is "SDN routing", which uses Software Defined Network (SDN) for route control. The second is "Passive selection", which is used to select incoming Proxy CAM packets. Furthermore, we will discuss the advantages and disadvantages of these methods and the future plan of this ongoing work.**

Keywords–CITS; Proxy CAM; Wide transmission; Grid Proxy CAM; SDN.

I. Introduction

Interest in autonomous driving has increased recently. Autonomous driving can be divided into several categories, according to function, from Level 0 to Level 5.

- Level 0 (No Automation): the vehicle is unassisted by Artificial Intelligence (AI). Autonomous driving is unavailable.
- Level 1 (Driver Assistance): AI performs only one driving operation, such as steering, acceleration, or deceleration. Automatic braking and adaptive cruise control are examples.
- Level 2 (Partial Automation): AI can do several driving operations simultaneously. Drivers must keep control and monitor the AI's decisions. In Japan, a system called "Pro Pilot" [1] by Nissan accelerates, decelerates, and steers to maintain the inter-vehicular distance and keep to the center of the road by detecting the vehicle in front and road lines.
- Level 3 (Conditional Automation): AI performs every operation of driving. However, if there is an emergency, the AI notifies the driver who takes over. Additionally, there are limitations to the driving environment, such as only highways.
- Level 4 (High Automation): AI does every operation of driving, even during an emergency. However, there is a limitation of the environment.
- Level 5 (Full Automation): AI does every operation in all circumstances. The driver does not need to keep watch. All passengers are free from driving.

For now, Levels 1 and 2 are commercialized, and Levels 3 and 4 are realized only at the research level. Intelligent Transport Systems (ITSs) manage problems related to an accident, traffic jam, environmental pollution, and the pursuit of convenience and comfort in road traffic. Autonomous driving is a key ITS technology, and the main research field is now stand-alone.

ITSs use only the vehicle's own sensors to make the next move. However, at an intersection, there is always a blind spot (i.e., an area the sensors cannot detect), and this often contributes to accidents. Cooperative ITS (CITS) [2] is a system that utilizes the communication among vehicles and roadside units (RSU) to share road traffic information and make blind spots smaller. Additionally, because this system can get the information from a distance, vehicles can change paths to avoid traffic jams, thus increasing convenience.

There are problems with designing the CITS network among vehicles and RSUs because of the communications protocols and the wireless technologies needed to create a real-time control system. Vehicles move very fast. Thus, vehicle information has a very short life span. This causes each vehicle to broadcast the information at a high frequency. The delay between sending and receiving also must be very short. The active distribution and balancing of network traffic is also very important. IEEE 802.11p [3] is a wireless technology for connecting vehicles, but it has a low fault tolerance and a short range, which must be improved for CITS.

In this work, we propose Grid Proxy Cooperative Awareness Message, to solve the problem of dispersing network traffic load and to improve the low fault tolerance and short range of IEEE 802.11p. There are two methods to realize this system: Software Defined Network (SDN) routing and passive selection. We describe these methods and discuss their advantages and disadvantages in this paper. By leveraging our previous work, Remote Proxy CAM [4], we improve the safety of Grid Proxy CAM.

The rest of this paper is organized as follows. Section II highlights CITS in detail, including protocol stack and two types of messages. Section III discusses related works, such as SDN, routing control, and Vehicular Ad-hoc Network (VANET). Additionally, we introduce works that have attempted to solve SDN's problems. In Section IV, we describe IEEE 802.11p's problems of network traffic load, low fault tolerance, and short range. In Section V, we finally define Grid Proxy CAM to solve the problems. Section VI concludes our paper and presents our future studies.

II. CITS

CITS is being designed to make road traffic more convenient and safer. The stand-alone system uses only the information from the sensors attached to vehicles to assist the driver's operation. However, CITS also enables vehicles to communicate with vehicles and RSUs and share information about road traffic. With this combined information, CITS-enabled vehicles can serve as better driving assistants and make better decisions. There are CITS architectures designed separately in Europe, America, and Japan. In this work, we focus on European CITS architecture.

CITS is composed of four primary layers: application, facility, network & transport, and access. There is also a management layer that manages the facility, network & transport, and access layers and a security layer that is responsible for CITS safety.

The access layer manages the wireless technologies (e.g., Long Term Evolution (LTE) and IEEE 802.11p [3]) and the network & transport layer manages the route control (e.g., GeoNetworking [5] and Basic Transport Protocol [6]). The facility layer summarizes and stores the information from sensors and communication messages to make it easier to utilize for applications, such as local dynamic mapping and other services.

There are two types of CITS communication massages: CAMs [7] and decentralized environmental notification messages (DENM) [8]. CAM is a message that every CITS-enabled vehicle and pedestrian cellphone generates. It contains highly dynamic information. With this type of message, CITS devices recognize the senders ' location and vector. Because of CAMs ' time sensitivity, messages must be generated at a high frequency, such as the recommended 1-to-10 Hz. CAM information may affect a small range (i.e., 100 m), and to mitigate CITS network bandwidth, messages are single-hop broadcasted. Thus, CAM message forwarding is not recommended. However, DENMs are messages that contain relatively static information about events that affect road and traffic conditions, such as construction, accidents, and weather. There is no recommendation for the frequency of transmission. So, CITS devices transmit DENMs according to their own configurations. DENM is recommended to multi-hop because the information it contains is considered to affect a wide range. When a CITS device receives CAMs and DENMs, they are delivered to facility layer and processed to provide the information to the application layer.

III. RELATED WORKS

A. SDN

In legacy networks, protocols, such as Open Shortest Path First (OSPF) and Routing Information Protocol (RIP) perform routing control according to certain algorithms. Special route configuration for each node must also be performed manually. However, if there is a replacement of nodes or a change in network topology, the configuration must be also performed manually. They cost a lot to maintain. However, with SDNs, a controller manages network routing. Thus, it does not need to be configured manually. In legacy networks, route control and data transmission is performed on the same network. In SDN, it is done on two separate networks: a control plane and a data plane. The control plane is composed of switches and a server (i.e., controller). This plane is used for sending control messages for routing and switch configuration, and for sending "packet-in" messages to notify the existence of a packet that does not match any routing configuration. The data plane is composed of switches and nodes. This network is used for transmitting and receiving data packets. In this work, we use Open Flow [9], a technology that supports SDN.

In Open Flow, routing configurations are called flow entries and are stored in a flow table. Flow entries are composed of rule and action fields. The rule field includes a condition; the switch checks if an incoming packet satisfies the condition. The action field includes an operation; the switch performs the operation if the packet meets the condition. The rule and action fields can be described by layers 1~4 of the Open Systems Interface reference model. If the controller modifies switch flow entries, it sends a "Flow-Mod" message to the switch. Then, whenever the switch receives a packet, it checks the flow table and performs the appropriate operation. If no flow entry requirement is met, the switch drops the packet or sends a "Packet-In" message to the controller. Then, the controller responds with "Packet-Out."

B. VANET

The Vehicular Ad-hoc network (VANET) is a system that enables CITS-enabled vehicles to form an ad-hoc network and communicate. All vehicles must perform routing control because of the ad hoc nature. The wireless technology used in VANET is IEEE 802.11p [3]. It is designed to enable fast-moving nodes to communicate, and it uses a high-frequency band (i.e., 5.9 GHz) because the distance between the nodes changes very fast and the time during which nodes remain within the communication range is very short; hence the speed of data communication must be fast.

One of the problems associated with VANETs is the occupancy of the communication bandwidth by periodic packets, such as CAMs. The higher the frequency of CAM transmission, the safer the road traffic becomes. However, increasing the frequency also increases the bandwidth occupancy due to higher data traffic. Hence, in a urban areas where the vehicle density is high, the transmission of CAMs exceeds the communication bandwidth and the packets that are transmitted when certain events occur, such as DENMs cannot be transmitted. Additionally, delay and packet loss can occur to the CAM packets themselves.

Another problem is forwarding the packets that need to be broadcasted and forwarded, such as DENMs. Basically, a node that receives such a packet must forward the packet to all nodes that are connected to the node. However, to applying this operation to all nodes could create forwarding loops and a large number of transmissions over a short duration. These transmissions could occupy the communication bandwidth and cause a long delay in the delivery of other messages. This delay is critical to the messages that require real-time property and this must be avoided.

For the first problem associated with VANETs, [10] concluded that nodes establish more connections than needed, making the transmission burden larger. Thus, they proposed fair-power adjustment for vehicular environments to control the network load of periodic packets (e.g., CAM) by setting an upper limit to network connections (i.e., Max Beaconing Load (MBL)). This forbids the number of connections at each node from exceeding the MBL. To adjust the number of the connections, they utilized the transmit power of wireless technologies. Their method avoided packet loss and delay. Additionally, if multi-hop packets (e.g., DENM) arrived to the network, they were less likely to not be broadcasted.

In [11], periodic packets constituted large data traffic because of the unnecessarily high frequency of transmission. Thus, the researchers proposed an adaptive beaconing rate to adjust the frequency of the transmission according to the surrounding situation and the node's status. They fed vehicle status (i.e., accident status and likelihood of causing an accident) and percentage of same-directional neighbor vehicles (i.e., density of vehicles moving in the same direction) into a fuzzy inference engine to calculate the precise transmission frequency needed for the situation.

For the second problem, [12] proposed The Last One (TLO) algorithm, which forces only the most appropriate nodes to broadcast packets. The algorithm transmits backwards from the sender, which is useful for mitigating an accident. Using Global Positioning System (GPS) information and the predefined distance of the communication range, each vehicle decides whether it is the Last Vehicle (LV) that received the packets from a sender. If so, it broadcasts the packets. To describe this algorithm in detail, if a node receives multi-hop broadcast packets, it receives the GPS information of the sender from the packets and calculates the geographical distance between the sender and the node. Similarly, the node calculates the geographical distance among the surrounding nodes. These nodes' GPS information is delivered from their CAMs. The communication range is common to all nodes and never changes. Thus, the node checks if there is another node behind the transmitting node within the communication range of the sender. If so, the node does not broadcast the packets for a while. If another node does not broadcast the packets and the node does not receive the packets, it concludes it is the LV and broadcasts the packets. Also, if the node cannot find any node that satisfies the condition, it decides that it is the LV and broadcasts the packets. In [13], the authors attempted to solve this problem by using the cellular network. They divided vehicles into two types: Gateway service Providers (GP), which communicate via IEEE 802.11p and cellular wireless technology; and ordinal vehicles, which communicate only via IEEE 802.11p. The periodic packets are assumed effective in for short range from the origin. Thus, they are broadcasted locally via IEEE 802.11p. Multi-hop packets are assumed to have a wider effect. Thus, when they are received by GPs, they send the packets to the cloud server. The server gets the GPS information of the origin from the packets and identifies the area needing the packets to be disseminated and sends them to GPs in that area. When the GPs receive the packets, they rebroadcast them.

C. *VANET using SDN*

There are many works to solve the problem of the occupancy of the communication bandwidth by massive packets. Recently, some methods have proposed using SDN for dynamic route control.

In [14], vehicles joined both VANET and cellular network and RSUs opened ad-hoc connection with vehicles. Additionally, the SDN controller was connected to RSUs and cellular base stations. The connection between SDN controller and RSUs and the SDN controller and vehicles belongs to control plane and the connection between RSUs and vehicles and vehicles each other belongs to data plane. RSUs and vehicles send information of connected RSUs and vehicles periodically to the SDN Controller via control plane. The SDN controller alters RSUs and vehicles' flow entries according to this information and when some specific nodes send requests to the SDN controller. Also, since vehicles and the SDN controller are connected via wireless technology, the connection can be lost. In that case, each vehicle has a local SDN agent. This agent performs the routing control using ordinal VANET routing protocols (e.g., Greedy Perimeter Stateless Routing (GPSR), Ad-hoc On-Demand Distance Vector (AODV), Destination-Sequenced Distance Vector (DSDV), Optimized Link State Routing (OLSR)) when the connection to SDN controller is lost.

In [15], the authors proposed the SDN-based Geographic Routing (SDGR) protocol for route control. The network topology is same as in [14]. This protocol is designed to be used when a vehicle sends messages to another vehicle. First, the sender only knows the IP address of the target, but does not know its geographical position. Thus, the sender sends a request to the SDN controller. When the SDN controller receives the request, it uses an optimal forwarding path algorithm. First, the controller identifies the geographical position of the target by using the information from periodic RSUs and vehicles messages sent to the controller. After that, the controller chooses the geographical path that goes through the relatively high density of vehicles in order to not lose the packets and to obtain the shortest path. After choosing the path, the controller replies with the path choice called optimal forwarding path (ofp) to the sender. When the sender receives the ofp, it inserts it into the packets. Using the packet forwarding algorithm and the ofp, the sender selects the next node and transmits the packets. The next node checks the ofp in the packets and uses packet forwarding algorithm to transmit the packets. All nodes in the path do the same operation, and eventually the packets reach the target. In SDGR, there are two modes: forthright mode and junction mode. The forthright mode is used when a vehicle is not at an intersection. When a vehicle is in this mode, it only checks the ofp of the packets and forwards them to the node which is on the path of the ofp, nearest to the target, and connected to the vehicle. When a vehicle is in junction mode, it compares its buffer occupancy with threshold δ. If the buffer occupancy is higher than δ, it broadcasts an Alarm message. When surrounding vehicles receive the message, they ignore the vehicle when using packet forwarding algorithm.

D. *Proxy CAM*

In CITS, a CITS enabled vehicle can detect vehicles which transmit CAMs and can be captured by the vehicle's sensors. However, IEEE 802.11p is weak to obstacles and it can be blocked by buildings easily. This creates a type of blind spot that no vehicle can be detected by either the sensors or CAMs, which can contribute to an accident. Additionally, there are vehicles that are not CITS-enabled. These vehicles need to be detectable by CITS-enabled vehicles. To solve this problem, [16] proposed Proxy CAM system, which installs computer vision sensors at the roadside and leveraging images captured by them. A server detects the vehicles by the images and makes CAMs for them and broadcasts them from transmitters installed at the roadside using IEEE 802.11p. In our previous work, we proposed Remote Proxy CAM [4], which delivers CAMs over the Internet using UDP/IPv6 and LTE with standard specification (i.e., basic transportation protocol/geonetworking and IEEE 802.11p).

IV. Problem Statement

In this section, we analyze the problems of Proxy CAM system. There are four problems: short wireless range, low fault tolerance, inefficient routing, and tradeoff between wide transmission and traffic load. We describe them in detail below.

A. *Short wireless range*

In [17], the Packet Delivery Ratio (PDR) of a vehicle traveling at 20 km/h using IEEE 802.11p with a data rate is 12 MB/s was nearly 100% when the range was to 700 m. However, this experiment was done in the flat plains with no buildings around and nearly no other wireless communication. However, in urban areas, there are buildings and other wireless communication. This environment can cause multipath propagation and wave interference. Thus, in this environment, the wireless range of IEEE 802.11p will diminish. Also, this Proxy CAMs can be used to determine the density of intersections. This information is very valuable to many applications, such as navigation for avoiding traffic jams. So, Proxy CAMs need to be provided enough early and this makes the problem of how to make the transmission far enough.

B. *Low fault tolerance*

IEEE 802.11p uses a relatively high-frequency bandwidth, 5.9 GHz. And this makes it weak to obstacles. Proxy CAM system overcomes this problem by installing transmitters around intersections. However, if the road connected to the intersection is curvilinear or if there is a large track, it may interrupt the communication between the Proxy CAM device and a vehicle. This interruption may be temporal, but for safety, it is a critical hazard.

C. *Inefficient routing*

Described in short wireless range, there is an unavoidable case that a vehicle needs far Proxy CAMs and cannot get them because of the communication range of IEEE 802.11p. To solve this problem, one solution is to relay the Proxy CAM packets. The devices to relay and broadcast the packets (i.e., transmitters) can be vehicles or RSUs. This relaying the packets adds the problem of how to perform the routing control. This routing control includes identifying devices for relaying, and how far Proxy CAM packets need to be relayed. Additionally, there are road traffic-related factors, such as traffic volume, accident rate in the past, and time-of-day. This routing must consider these factors and provide alternate routing for each Proxy CAM device.

D. *Trade-off between wide transmission and traffic load*

Described before, it needs to lengthen the communication range of Proxy CAMs to improve the safety. And to do this, relaying Proxy CAM packets and transmitting from a remote transmitter is the solution. However, for the transmitter, if the range is too long, the number of the Proxy CAM devices that use this transmitter increases dramatically. This means the number of the Proxy CAM packets also increases. For wired communication, the amount of the traffic load is small, but for wireless communication, especially IEEE 802.11p, the bit rate is 3∼27 MB/s and this is used for broadcasting Proxy CAM packets to vehicles in Proxy CAM system. Thus, if the amount of Proxy CAM packets is too high, the transmitter cannot handle the packets and this makes packet loss and delay.

V. Grid Proxy CAM

To solve the problems in the previous chapter, we propose Grid Proxy CAM system. This system is basically composed of Proxy CAM device and relaying devices. In inefficient routing problem described in the previous chapter, this problem would be solved by relaying the packets, and devices to relay the Proxy CAM packets can be vehicles or RSUs. In [15], they used vehicles. Unfortunately, this method depended on vehicle speed and density, and these factors change quickly in the real world. The quality of safety-related services must always be high, and the extant proposals were not suitable for relaying Proxy CAM packets. So in this study, we use RSUs as the relaying devices that are routers installed at each intersection with a Proxy CAM device. For the problems of inefficient routing and the trade-off between wide transmission and traffic load, we have been developing two methods: SDN routing and Passive selection.

A. *SDN routing*

Figure 1 shows the overview of SDN routing method. This method uses SDN for routing, and all routers relay the Proxy CAM packets by following their SDN flow table. Each Proxy CAM device is composed of a Proxy CAM generator that detects surrounding vehicles with computer vision and generates their Proxy CAMs, and a Proxy CAM transmitter that broadcasts Proxy CAMs using IEEE 802.11p. These generators and transmitters are connected to each router, and adjacent routers are also connected. Each router connected to an SDN controller. These connection are wired, and all Proxy CAM generators, transmitters, routers, and the SDN controller have an IP address and communication between generators, transmitters, and routers use UDP.

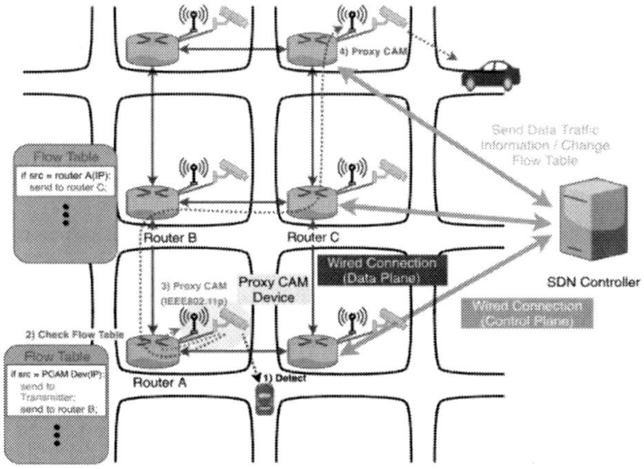

Figure 1. Overview of Grid Proxy CAM with SDN routing.

The SDN controller changes the flow entries of each router's flow table periodically according to the traffic load of each transmitter which is connected to the router, and the priority. This priority indicates how important Proxy CAMs generated by the directly connected Proxy CAM generator are to road safety. The factors for determining this priority includes the type of the street where the Proxy CAM device is installed (i.e., the street is the main street or not), the accident rate in the past, whether the street is curvilinear and it is difficult to see far, the density of vehicles, whether there is a fast moving vehicle that can cause an accident, and whether there is an emergency vehicle. With the traffic load and the priority, the SDN controller changes the flow entries so that each transmitter can broadcast as many Proxy CAM packets as possible. However, their amount does not exceed the capacity of the transmitter's IEEE 802.11p limitations.

The procedure is as follows.

1) A Proxy CAM generator detects surrounding vehicles and generates Proxy CAMs and sends to a connected router.
2) When the router receives the packets, it checks its SDN flow table and follows instructions. By default, in this flow table, there is an entry that checks if the Proxy CAM packets are from the directly connected Proxy CAM generator, sending them to the directly connected transmitter. Additionally, the SDN controller can add and delete flow entries to duplicate the packets and sends them to adjacent routers.
3) The directly connected transmitter broadcasts the packets with IEEE 802.11p as soon as it receives the packets.
4) The router also has the flow entries for incoming packets from neighbor routers, used to duplicate the packets and send to adjacent routers and the directly connected transmitter.

During this procedure, the SDN controller always gets necessary information from routers and periodically changes flow entries of each router.

B. *Passive selection*

Figure 2 shows the overview of passive selection method. The hardware architecture of this method is very similar to SDN routing. There are routers and Proxy CAM generators and Proxy CAM transmitters. Each Proxy CAM generator and Proxy CAM transmitter is connected to a router, and the router is connected to adjacent routers. These connections are wired and use UDP/IP protocol. Transmitters broadcast Proxy CAM packets using IEEE 802.11p. In this method, we assume that communication speed of wired connections is enough high and the data load of Proxy CAM packets does not occupy the bandwidth. Each router has a geographical range and it receives Proxy CAM packets from routers which are within the range. In detail, each router has a list of other routers' IP addresses which are within the range in advance, and the routing is done with RIP or OSPF, which are legacy network's protocols. With the IP address list, each router sends Proxy CAM packets.

When a router receives Proxy CAM packets, it duplicates them and sends to the next routers and to a directly connected transmitter. The transmitter checks the number of packets. If it is less than the bandwidth of IEEE 802.11p, it broadcasts all the packets using IEEE 802.11p. However, if the number is

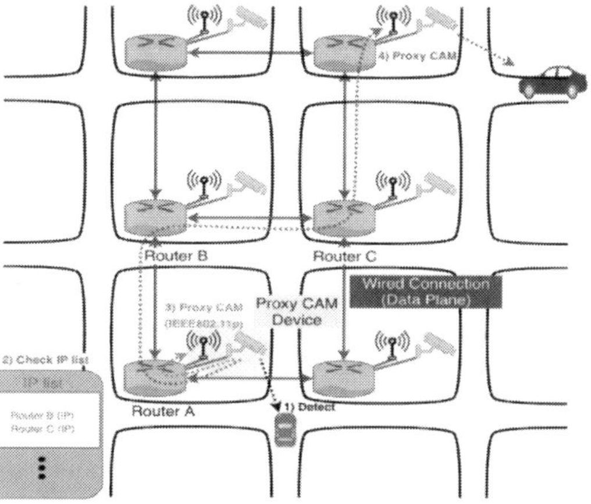

Figure 2. Overview of Grid Proxy CAM with Passive selection.

more than the bandwidth, it is impossible to send all of them. Thus, the transmitter selects the Proxy CAM and broadcasts the selected packets. The factors for how to select the packets would be the geographical distance to the original Proxy CAM device, the street on which the original Proxy CAM device is installed is the main street or not, the street is curvilinear or not, the density of vehicles, if there is a fast-moving vehicle, and if there is an emergency vehicle.

The procedure is as follows.

1) A Proxy CAM generator detects the surrounding vehicles and generates Proxy CAMs and sends to a connected router.
2) When the router receives the packets, it checks its IP list and sends the packets to the routers listed. The routes to them are established using the original network protocols. Additionally, it sends the packets to the directly connected transmitter.
3) The directly connected transmitter broadcasts the packets using IEEE 802.11p as soon as it receives the packets.
4) The transmitter also broadcasts the packets from other Proxy CAM generators. However, if the data load of the packets exceeds the capacity of IEEE 802.11p, it selects the packets and drops the rest.

VI. Conclusions and Future Work

In this paper, we proposed Grid Proxy CAM system that uses RSUs as routers. Each router is connected to Proxy CAM generator, transmitter, and also neighboring routers. We introduced two methods of SDN routing and passive selection that forward Proxy CAM packets and control their load.

SDN routing makes the route control easy and dynamic. This makes the system flexible to network and situational changes (e.g., when a new road is built or an existing road is extended). Additionally, all routes are defined by an SDN controller and the controller defines only the necessary routes. Thus, the load to not only IEEE 802.11p that transmitters use but also every wired connection in this system should be minimum. In addition to these points, routers and transmitters are not given some heavy task to handle the packets. This means their performance can be low. Thus, the price for the system can be low. On the other hand, SDN routing is done by an SDN controller. This routing process can still be heavy. Thus, the controller must be high-performance. Additionally, if a failure occurs to the controller, all routers cannot change the routes, which may cause the system to stop.

Passive selection is a decentralized system. This means there is no single point of failure. Thus, this system is strong. Additionally, this method does not require complex tasks and uses legacy network technology. Thus, it is easy to install. On the other hand, passive selection is not as flexible as SDN routing. It requires manual operation to change IP address lists and the range of getting Proxy CAM. Additionally, this system requires transmitters to select the packets. If this process takes time, the property of real-time may be lost.

We are also considering that our previous work, Remote Proxy CAM [4] can be combined with this Grid Proxy CAM. The remote Proxy CAM system is composed of Proxy CAM devices and a server. Vehicles can access the server by sending a request with their position via LTE and can get Proxy CAM information. When the server receives a request, it checks the position of the source vehicle and gathers the Proxy CAM from routers that are within a certain range from the vehicle. It then sends them to the vehicle. In this system, vehicles use LTE to communicate, and this wireless technology covers almost every location in the city. Thus, theoretically, vehicles can get all information from Proxy CAM devices in the city. With this different characteristic, we are considering Grid Remote Proxy CAM system that uses Grid Proxy CAM system for collecting nearby Proxy CAMs and Remote Proxy CAM for collecting remote Proxy CAMs. This combined system will use both IEEE 802.11p and LTE. This means this system disperses the network traffic load and is also strong to radio wave interference.

For future work, we plan to propose algorithms of SDN controller's altering flow entries and passive selection's selecting incoming packets. Additionally, we will implement this system in a network simulator and perform experiments. With the outcome of these experiments, we will discuss the effect of the methods. In addition to this, we will implement the combined Grid Remote Proxy CAM system and examine how much this system improves safety.

Acknowledgement

This work was partly supported by JSPS KAKENHI Grant Number JP17H04678.

References

[1] ProPILOT, retrieved: 06, 2018. [Online]. Available: https://www.nissan-global.com/JP/TECHNOLOGY/OVERVIEW/propilot.html

[2] ETSI, "TR 102 962 - V1.1.1 - Intelligent Transport Systems (ITS); Framework for Public Mobile Networks in Cooperative ITS (C-ITS)," 2012, retrieved: 06, 2018. [Online]. Available: http://www.etsi.org/deliver/etsi_tr/102900_102999/102962/01.01.01_60/tr_102962v010101p.pdf

[3] IEEE Computer Society. LAN/MAN Standards Committee., Institute of Electrical and Electronics Engineers., and IEEE-SA Standards Board., IEEE standard for Information technology– telecommunications and information exchange between systems– local and metropolitan area networks– specific requirements : Part 11 : Wireless LAN medium access control (MAC) and physical layer (PHY) specifications : Amendment 6: Wireless access in vehicular environments. Institute of Electrical and Electronics Engineers, 2010, retrieved: 06, 2018. [Online]. Available: http://ieeexplore.ieee.org/document/5514475/

[4] M. Kitazawa, M. Tsukada, K. Morino, H. Ochiai, and H. Esaki, "Remote Proxy V2V Messaging using IPv6 and GeoNetworking," in The Sixth International Conference on Advances in Vehicular Systems, Technologies and Applications (VEHICULAR 2017), July 2017, pp. 74–80, retrieved: 06, 2018. [Online]. Available: https://hal.archives-ouvertes.fr/hal-01578410

[5] Its, "EN 302 636-5-1 - V1.2.0 - Intelligent Transport Systems (ITS); Vehicular Communications; GeoNetworking; Part 5: Transport Protocols; Sub-part 1: Basic Transport Protocol," retrieved: 06, 2018. [Online]. Available: http://www.etsi.org/deliver/etsi_en/302600_302699/3026360501/01.02.00_20/en_3026360501v010200a.pdf

[6] ——, "TS 102 636-4-1 - V1.1.1 - Intelligent Transport Systems (ITS); Vehicular communications; GeoNetworking; Part 4: Geographical addressing and forwarding for point-to-point and point-to-multipoint communications; Sub-part 1: Media-Independent Functionality," retrieved: 06, 2018. [Online]. Available: http://www.etsi.org/deliver/etsi_ts/102600_102699/1026360401/01.01.01_60/ts_1026360401v010101p.pdf

[7] ETSI, "EN 302 637-2 - V1.3.1 - Intelligent Transport Systems (ITS); Vehicular Communications; Basic Set of Applications; Part 2: Specification of Cooperative Awareness Basic Service," retrieved: 06, 2018. [Online]. Available: http://www.etsi.org/deliver/etsi_en/302600_302699/30263702/01.03.02_60/en_30263702v010302p.pdf

[8] ETSI EN 302 637-3 v1.2.2(2014-11), Intelligent Transport Systems(ITS); Vehicular Communications; Basic Set of Applications; Part 3: Specifications of Decentralized Environmental Notification Basic Service, 2014, retrieved: 06, 2018. [Online]. Available: http://www.etsi.org/deliver/etsi_en/302600_302699/30263703/01.02.02_60/en_30263703v010202p.pdf

[9] Software-Defined Networking (SDN) Definition, retrieved: 06, 2018. [Online]. Available: https://www.opennetworking.org/sdn-definition/

[10] M. Torrent-Moreno, P. Santi, and H. Hartenstein, "Fair sharing of bandwidth in VANETs," in Proceedings of the 2nd ACM international workshop on Vehicular ad hoc networks - VANET '05. New York, New York, USA: ACM Press, 2005, p. 49.

[11] K. Zrar Ghafoor, K. AbuBakar, M. van Eenennaam, R. H. Khokhar, and A. J. Gonzalez, "A fuzzy logic approach to beaconing for vehicular ad hoc networks," Telecommunication Systems, vol. 52, no. 1, January 2013, pp. 139–149.

[12] K. Suriyapaibonwattana and C. Pomavalai, "An Effective Safety Alert Broadcast Algorithm for VANET," in 2008 International Symposium on Communications and Information Technologies. IEEE, October 2008, pp. 247–250.

[13] B. Liu, D. Jia, J. Wang, K. Lu, and L. Wu, "Cloud-Assisted Safety Message Dissemination in VANETCellular Heterogeneous Wireless Network," IEEE Systems Journal, vol. 11, no. 1, March 2017, pp. 128–139.

[14] I. Ku et al., "Towards software-defined VANET: Architecture and services," in 2014 13th Annual Mediterranean Ad Hoc Networking Workshop (MED-HOC-NET). IEEE, June 2014, pp. 103–110.

[15] X. Ji, H. Yu, G. Fan, and W. Fu, "SDGR: An SDN-Based Geographic Routing Protocol for VANET," in 2016 IEEE International Conference on Internet of Things (iThings) and IEEE Green Computing and Communications (GreenCom) and IEEE Cyber, Physical and Social Computing (CPSCom) and IEEE Smart Data (SmartData). IEEE, December 2016, pp. 276–281.

[16] T. Kitazato, M. Tsukada, H. Ochiai, and H. Esaki, "Proxy cooperative awareness message: an infrastructure-assisted v2v messaging," in 2016 Ninth International Conference on Mobile Computing and Ubiquitous Networking (ICMU), October 2016, pp. 1–6, retrieved: 06, 2018. [Online]. Available: http://ieeexplore.ieee.org/document/7742092/

[17] O. Shagdar, M. Tsukada, M. Kakiuchi, T. Toukabri, and T. Ernst, "Experimentation towards ipv6 over ieee 802.11p with its station architecture," in International Workshop on IPv6-based Vehicular Networks, June 2012, pp. 1–6, retrieved: 06, 2018. [Online]. Available: https://hal.archives-ouvertes.fr/hal-00702923

Evaluation of WiFi Access Point Switching for Vehicular Communication Using SDN

Kaito Iwatsuki
Computer and Information Science
Graduate School of
Science and Engineering
Doshisha University
Kyoto, Japan
email:kaito.iwatsuki@nislab.doshisha.ac.jp

Nishiki Hase
Computer and Information Science
Graduate School of
Science and Engineering
Doshisha University
Kyoto, Japan
email:nishiki.hase@nislab.doshisha.ac.jp

Kenya Sato
Computer and Information Science
Graduate School of
Science and Engineering
Doshisha University
Kyoto, Japan
email:ksato@mail.doshisha.ac.jp

Abstract—**The amount of data required by various applications continues to increase due to improved high function terminals. We expect that using Long Term Evolution (LTE) networks will also grow in the future. Networks might become very crowded depending on the places and the times at which people congregate. To reduce network congestion, carriers are offloading LTE network traffic to WiFi systems. Because the frequency resources of LTE networks are limited, movements that utilize them will increase in the future. When a mobile terminal communicates using a communication system with a narrow coverage area, such as WiFi, the frequency of switching the Access Point (AP) increases compared to using Cellular Networks. Since AP switching frequently occurs, network connections must be made more quickly. Here, the communication disconnection time denotes an efficient switching. Therefore, in this research, we select an AP for connections based on position and speed, both of which are unique to mobile terminals, and utilize the concept of a Software Defined Network (SDN). We propose a method that enables efficient network connections by executing an AP procedure that is connected before the communication is disconnected.**

Keywords–vehicular communication; Wi-Fi; SDN.

I. INTRODUCTION

In recent years, communication is often done through many mobile terminals including smartphones. Such Cellular Networks as LTE are mainly used for these mobile units. Since Cellular Networks are becoming congested by an increase in the number of mobile terminals, such applications require more data. To reduce the congestion of Cellular Networks, each carrier is offloading its cellular network traffic to unlicensed bands, such as WiFi systems. In fact, according to a Cisco survey, 58% of vehicular communication is offloaded traffic [1], which is expected to continue to rise in the future. Therefore, the importance of WiFi systems for vehicular communication will also increase in the future. We also expect to utilize WiFi systems in communication even in such fields as automobiles. However, as mentioned above, when a WiFi system is used for vehicular communication, a new problem arises: the time during which communication cannot be performed becomes longer than with a cellular system when the Access Point (AP) of the WiFi system is switched for a connection with a mobile terminal. Compared to cellular systems, existing WiFi system have a smaller coverage area that includes just one AP, and the APs belonging to a plurality of different networks cooperatively lack a function for assisting the AP ' s terminal switch, for example. In other words, vehicular communication using WiFi has a higher disconnection frequency and a longer disconnection time than cellular systems. When vehicular communication is done using a WiFi system with a narrow AP coverage area, if such connection procedures as authentication take too much time, the mobile terminal leaves the coverage area before the authentication is completed. Therefore, in this research, we shortened the disconnection time of communication during AP switching. Among AP-switching operations when connecting to the AP, selecting the destination of the AP being switched and the authentication operation occupy most of the entire switching operation. In this research, we use the SDN (Software Defined Network) concept [2][3] on WiFi networks to shorten these two disconnection times to solve the problem of using WiFi systems in vehicular communication. The rest of the paper is organized as follows. Section II overviews related works and problems of conventional method. In section III, we describe the proposed method. Simulation results are provided in Section IV, and then Section V concludes the paper.

II. RELATED WORK

In this section, we describe the technologies that are related to WiFi vehicular communication, as well as current research on AP switching that occurs when using them.

A. *Problems with current WiFi switching*

When a mobile terminal communicates with a WiFi system, it must switch to the next AP as it leaves the area covered by each WiFi ' s AP. To switch to an AP, it must be disconnected from the system before it is connected to a new one. Then, a probe request is sent to the AP, as well as an authentication procedure.

B. *Disconnection from a Connected AP*

When the mobile terminal cannot receive a beacon signal that is transmitted every 100 ms by the WiFi's AP, it recognizes that communication with the AP has been disconnected and starts to scan for another AP. The allowable time when this beacon signal cannot be received varies by vendor and mounting method.

C. Disconnection from a Connected AP

AP information of WiFi, passive, and active scans.

- Passive scan
  The mobile terminal waits to receive a beacon signal from the AP to obtain its information.
- Active scan
  The mobile terminal transmits probe request information to the AP. Upon receiving this it, the AP describes its own information in the probe response and transmits it.

In either method, when switching the AP, we must disconnect from the AP to which we are currently connected and authenticate the next AP. Therefore, communication is disconnected until the authentication processing is completed. Another problem is that no communication can be performed when the mobile terminal leaves the coverage area without completing the authentication process within the coverage area. Although several methods have achieved the same purpose as our research, we just list two related researches. The first strengthens the cooperation between APs [4]. In this method, the network AP provides information on other APs, as well as itself, and the mobile terminal recognizes them in advance, shortening the disconnection time when this new AP is connected. With this method, only information on the AP in the LAN (to which it is connected) can be obtained. Therefore, this method cannot be applied when a mobile terminal switches APs that belong to different networks. The second method scans the AP while communicating with the connected AP and selects the AP's next destination to which it will be switched [5][6]. In this method, unfortunately, an AP, which is not yet in the communication range of the movement destination, cannot be considered a switching candidate because of a feature that scans the AP within its own communication range beforehand.

III. PROPOSAL

In this section, we propose a method to improve the efficiency of connection to the WiFi network.

A. Outline

In the proposed method, a mobile terminal selects the destination of the AP being switched in advance and executes the authentication procedure necessary for the connection. This shortens the communication disconnection time that occurs at the AP-switching time. Before communication is disconnected, AP candidates are not scanned directly by the mobile system but are selected based on the information obtained from the server that holds the AP information. The following is the general operation flow. First, the mobile terminal selects the next AP to be connected based on its own position and speed. Then, the mobile terminal requests the controller to perform the authentication process necessary for the next connection. In this way, our proposed method reduces the disconnection time during switching.

B. Precondition

In the implementation environment of this research, we assume that all the APs are compatible with SDN and that the AP, which is managed by the controller and such processes as authentication, can be performed based on instructions from the controller.

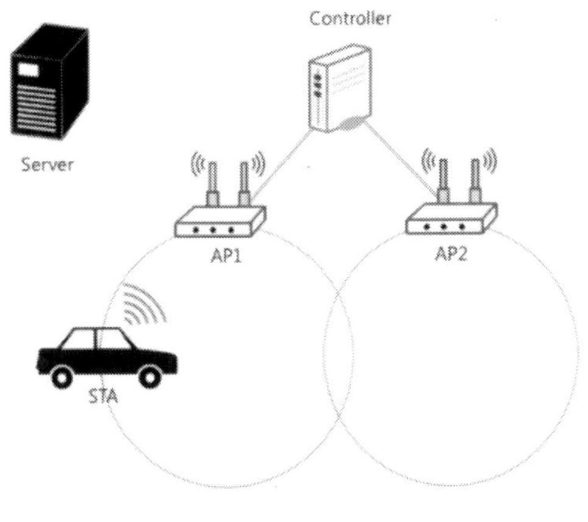

Figure 1. Component

C. Configuration

The configuration of our proposed method is shown in Figure 1.

- Mobile terminal (STA)
  The STA performs vehicular communication and obtains its own position information from GPS.
- Controller (controller)
  This controller is compatible with SDN and manages the APs in the network. When receiving an authentication request from an AP, it instructs the corresponding AP to perform authentication.
- Access point (AP)
  The AP is compatible with SDN and executes the processing when it receives an authentication instruction from the controller.
- Server (Server)
  It holds the AP's location information as well as the information of the controller that is managing the AP. When it receives a request from the STA, it provides information on the AP based on STA's position information.

D. Operating Sequence

The operation of the proposed method is roughly divided into two phases: pre-authentication and switching. Next, we describe their procedures.

E. Pre-authentication Phase

The flow of the pre-authentication phase is shown in Figure 2 and Figure 3.

1) The STA gets its position information.
2) It acquires the AP information in the vicinity by the connected AP.

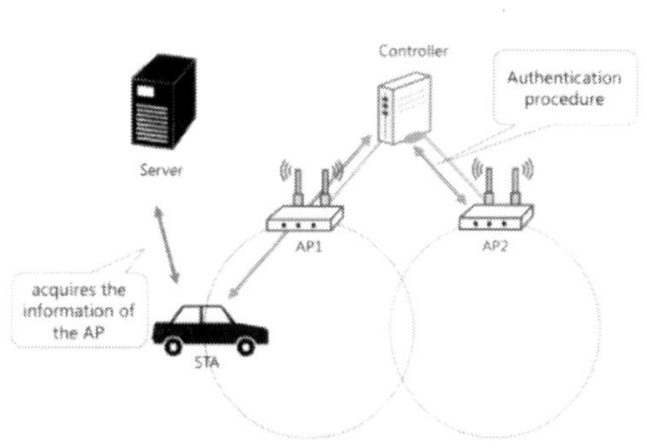

Figure 2. Acquires AP information

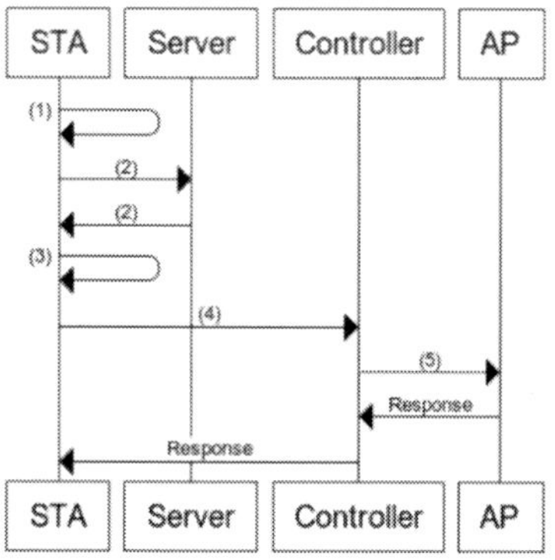

Figure 3. Sequence diagram of pre-authentication phase

3) From the server, it selects the next AP to be connected (the AP after switching) from its position and movement direction.

4) It requests controller authentication for the corresponding AP.

5) The controller sends an authentication instruction to the corresponding AP.

F. Switching Phase

The switching phase's flow is shown in Figure 4 and Figure 5. As a premise, the AP always distributes beacon signals to its surroundings.

1) The STA sends a probe request to the AP selected in the pre-authentication phase when the beacon signal times out the value specified by the Beacon Watchdog Timer. In the proposed method, its value is decreased only when the STA is fast, thereby shortening the communication disconnection time with the AP before switching.
2) After disconnection, the AP sends a probe response.
3) When the STA receives a probe response, it makes a connection by omitting an authentication because such an authentication procedure as a 4-way handshake (which are conventionally done) was carried out beforehand. Authentication procedures depend on the types of implementation. In this paper, we assume a four-way WPA2 handshake.

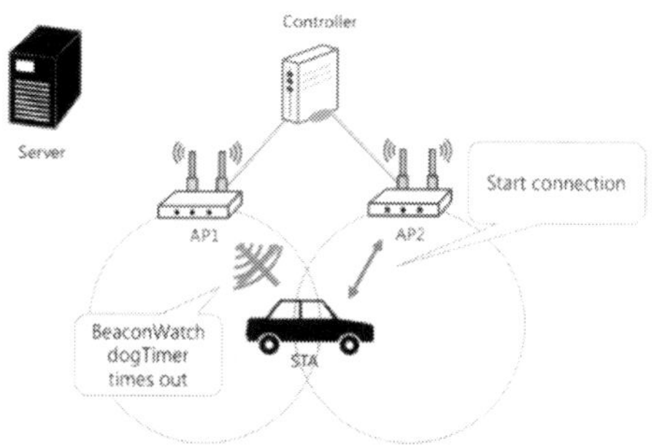

Figure 4. Switching phase

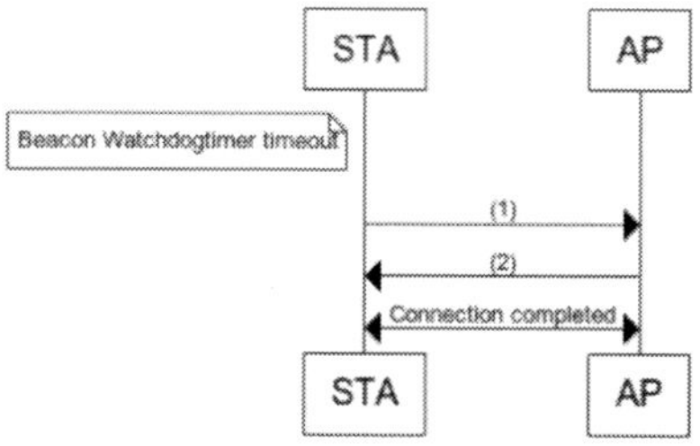

Figure 5. Sequence diagram of switching phase

IV. EVALUATION AND CONSIDERATION

A. Simulator

In this research, we used NS(Network Simulator)3 [7], which is a network simulator to evaluate our proposed method's performance. The algorithm of the proposed method was implemented in the data link layer and the network layer of the network node on NS3.

B. Assumed Use Case

We used various cases that involved vehicular communication. We assumed a car as a mobile terminal. Several usage scenarios are conceivable even for automobile usage, and they are roughly classified into the following two types:

- On the highway
  When used on a highway, the moving speed is fast, and the vehicle density is generally low.
- On general roads
  When used on an ordinary road, the moving speed is slow, and the car density is higher than when used on a highway.

Based on the above items, we made the following assumptions. One is using a cellular network when communicating on a highway with low vehicle density. Second, we used the proposed method when communicating on an ordinary road with a relatively high vehicle density.

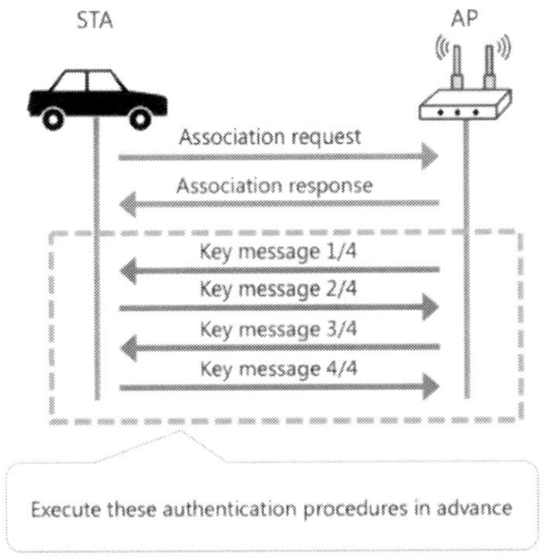

Figure 6. WPA2 key message exchange

C. Evaluation Environment

The evaluation environment is shown in Table I. Since NS3 does not have a default WiFi switching handoff function, we applied the ns3-wifi-infra-handoff patch and implemented the WiFi handoff function on NS3. We assume WPA2, which is currently the most secure authentication method, and exchange key messages among the procedures beforehand in the pre-authentication phase (Figure 6).

TABLE I. EVALUATION ENVIRONMENT

OS	Ubuntu 14.04
CPU	Intel core i7
Memory	8GB
Simulator	NS 3.22
Propagation delay model	ConstantSpeedPropagationDelayModel
Propagation loss model	LogDistancePropagationLossModel
Authentication method	WPA2

D. Evaluation Method

We modified the data link layer in the network node on NS 3 and virtually implemented our proposed method. The evaluation topology resembles that shown in Figure 7. The mobile terminal, which switches among a plurality of APs while moving. We measured it based on the disconnection time with the server that is communicating during the movement. The mobile terminal ' s speed is a constant 36 km/h, and the distance between each AP is 150 m. The disconnection time is the average of the time during which communication with the server (caused by one AP switching) cannot be performed. To measure the effect of each pre-authentication phase and switching phase, which are the two phases of the proposed method, we compared the results obtained by setting two different parameters.

1) Evaluation of Pre-authentication Phase: As described above, the evaluation criterion is the communication disconnection time during AP switching. We compared the evaluation results in the following two parameter settings.

- Conventional method

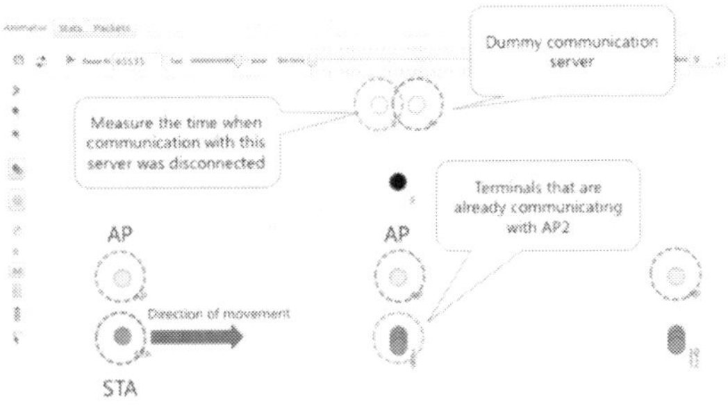

Figure 7. Evaluation topology

  In the conventional WiFi connection method, the Beacon Watchdog Timer uses the default value.
- Proposed method (only pre-authentication section implemented)
  The pre-authentication shown in the proposed method is done before switching, but the Beacon Watchdog Timer uses the default value. Only pre-authentication phase is implemented.

With these two comparisons, we measured the changes in the communication disconnection time due to the pre-authentication and its effect.

2) Evaluation of Switching Phase: The evaluation criteria are the same as the above pre-authentication phase, but we changed the parameters that were compared as follows and evaluated the switching phase:

- Proposed method (only pre-authentication section implemented)
  In the parameter settings, which were also used in the evaluation of the pre-authentication phase, we only implemented the pre-authentication part, and the Beacon Watchdog Timer uses the default value.
- Proposed method
  Evaluation when both of the two phases mentioned in the part of the proposed method are implemented.

We evaluated both above phases that were implemented with lowered disconnection times and measured the Beacon Watchdog Timer value and the switching phase effect. The topology in the actual simulator is shown in Figure 7. Apart from the elements described in the proposed section, we added terminals that have already been connected and are communicating with AP2. By adding this element, we created a more realistic environment and measured the changes in communication disconnection time by changing the numbers of this element.

E. Evaluation Result

Figure 8 shows the evaluation result of the pre-authentication phase, which is the transition of the disconnection time of the proposed and conventional methods. The x axis represents the number of terminals that are already connected, and the y axis represents the disconnection time. Both the conventional and proposed methods show that the disconnection time increases as the number of already connected

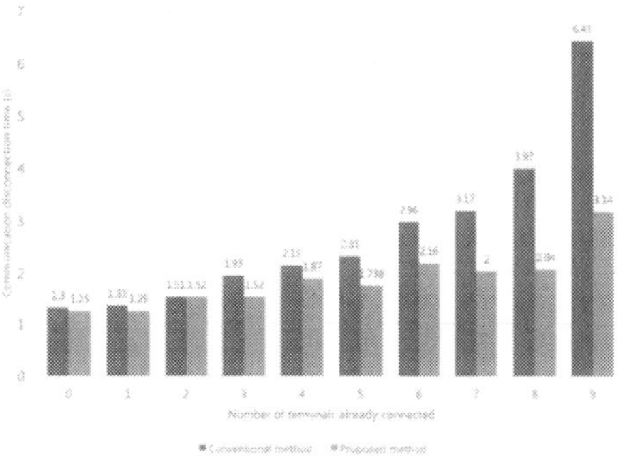

Figure 8. Evaluation result of pre-authentication phase

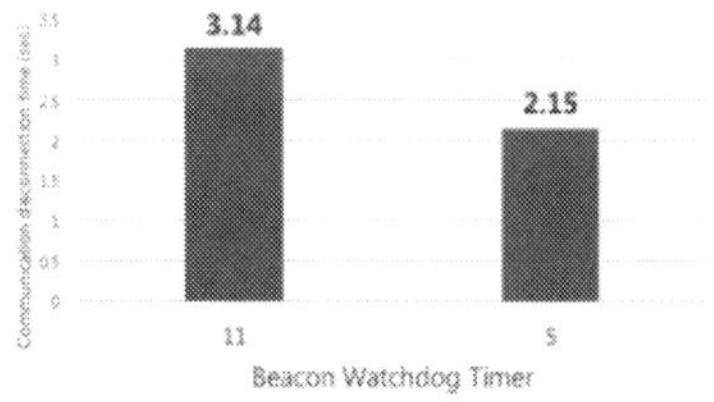

Figure 9. Evaluation result of switching phase

terminals increases. The increase in these disconnection times is probably affected by CSMA/CA, which was used in the WiFi system. Both methods show that the disconnection time increases in proportion to the increase of the number of already connected terminals, but the increase in the disconnection time of the proposed method is more gradual than the proposed method. Figure 9 shows the evaluation result of the switching phase. The number of already connected terminals is nine. In this case, STA had the longest cutting time when evaluating the pre-authentication phase. We compared the communication disconnection time when the Beacon Watchdog Timer value has a default value of 11 and the shortened value is 5. In this experiment, the AP transmits a beacon signal at a cycle of 100 ms, and the Beacon Watchdog Timer times out when it cannot receive a specified number of beacons.

F. Consideration of Evaluation Result of Pre-authentication Phase

As seen in Figure 8, which is the evaluation result of the pre-authentication phase, the disconnection time increases in both the conventional and proposed methods as the number of already connected terminals increases. However, when the number of already connected terminals is two or more, the communication disconnection time of the proposed method is lower than the conventional method. This is probably due to the reduction of the procedure that is done at the connection time by carrying out the necessary authentication procedure. In addition, as the number of already connected terminals increases, the difference in the disconnection time between the proposed and conventional methods increases. This result shows that the maximum bottleneck in the authentication process is the

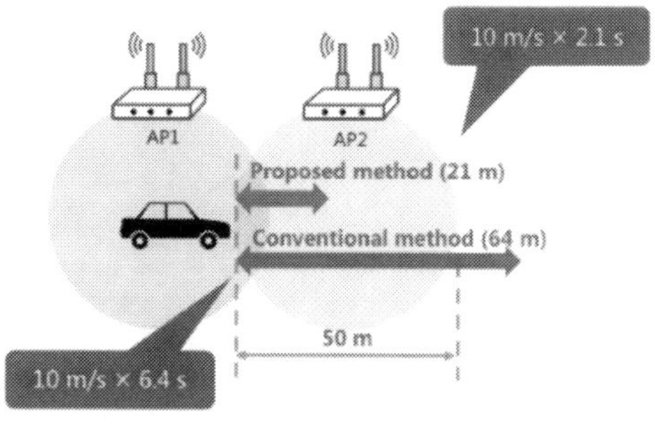

Figure 10. Distance required for authentication

waiting time of the CSMA/CA that is generated when the authentication frame is transmitted. When the number of already connected terminals increases, their communication might collide with each other, so CSMA/CA ' s back-off time increases. Since CSMA/CA collisions depend on timing, the evaluation result varies. As an example, the disconnection time with five terminals already connected is shorter than the disconnection time when the number of already connected terminals is four in the proposed method.

G. Consideration of Evaluation Result of Switching Phase

As seen from Figure 9, which is the evaluation result of the switching phase, the communication disconnection time can be shortened by reducing the Beacon Watchdog Timer value. This result is influenced by quickly disconnecting the connection with the AP that became unable to communicate and switching to the next AP. When carrying out vehicular communication, if the mobile system intends to maintain the AP connection for a long time and if the AP is actually already out of communicable range, it may lengthen the disconnection time. Our proposed method prevented this by lowering the Beacon Watchdog Timer value. However, decisive overhead exists in a technique that shortens this Beacon Watchdog Timer. If the radio waves temporarily deteriorate when the terminal is not moving, a mistaken disconnection will occur. In this research, we reduced this overhead by only shortening the Beacon Watchdog Timer when the mobile system has enough speed for this overhead.

H. Consideration of Authentication Failure

Even in the related research section, the problem is that authentication fails because the mobile terminal leaves the AP' s coverage area before the authentication process is completed, and so communication cannot be achieved. Regarding this problem, as shown in Figure 10 from the results of the pre-authentication and switching phases, the following idea can be addressed. Since the conventional method requires a maximum of 6.43 s for authentication, communication cannot be performed by exiting the coverage area before authentication is done in a general public WiFi coverage area of 50 m; authentication is completed while the proposed method can achieve a maximum of 2.15 s. That is, authentication has been completed within the coverage area and communication is achieved.

I. Timing of Switching

This time the connected terminal spontaneously determined the switching timing led by the Beacon Watchdog Timer. However, even in such a cellular network as an LTE, the communication base station determines the time of the terminal switching. By introducing this method to our proposed method, the superiority of the controller ' s centralized management by SDN may be improved.

V. CONCLUSION

In recent years, research movements to use WiFi systems for vehicular communication have been growing, and offloading traffic to the WiFi systems in mobile traffic will increase in the future. However, when using a WiFi system for vehicular communication, several problems arise. In this research, for the problem of mobile terminal communication that uses WiFi systems, we examined the communication disconnection time that occurs when APs are switched. We solved this problem by switching APs based on the location speed of the mobile terminal. Our proposed method is roughly divided into two phases. First, in the preliminary authentication phase, we performed the necessary authentication processing before establishing a connection with SDN technology. The second phase shortened the Beacon Watchdog Timer in the switching phase and switched the AP. Based on the evaluation results of both phases, we shortened the communication disconnection time by making switching more efficient. We also solved erroneous disconnections, which are the overhead considered in the switching phase, based on the mobile terminal ' s speed. Communication can be achieved using the proposed method even for APs that leave the coverage area before authentication is completed by a conventional method.

ACKNOWLEDGMENT

This work was supported in part by JSPS KAKENHI Grant Number 16H02814.

REFERENCES

[1] " Cisco Visual Networking Index: Global Mobile Data Traffic Forecast Update, 20152020 " , URL: https://www.cisco.com/c/dam/m/en_in/innovation/enterprise/assets/mobile-white-paper-c11-520862.pdf [accessed: 2017-10-10].

[2] D. Zhao, M. Zhu, and M. Xu, "SDWLAN: A flexible architecture of enterprise WLAN for client-unaware fast AP handoff", Fifth International Conference on Computing, Communications and Networking Technologies (ICCCNT), Hefei, 2014, pp. 1-6.

[3] M. Bernaschi, F. Cacace, and G. Iannello, "OpenCAPWAP,an open-source CAPWAP implementation for management and QoS support", 2008 4th International Telecommunication Networking Workshop on QoS in Multiservice IP Networks, Venice, 2008, pp. 72-77.

[4] S. Feirer and T. Sauter, "Seamless handover in industrial WLAN using IEEE 802.11k", 2017 IEEE 26th International Symposium on Industrial Electronics (ISIE), Edinburgh, 2017, pp. 1234-1239.

[5] P. Macha and J. Wozniak, "A lightweight algorithm for fast IEEE 802.11 handover", Australasian Telecommunication Networks and Applications Conference (ATNAC) 2012, Brisbane, QLD, 2012, pp. 1-6.

[6] D. Lee, D. Won, and M. J. Piran, "Reducing handover delays for seamless multimedia service in IEEE 802.11 networks", in Electronics Letters, vol. 50, no. 15, pp. 1100-1102, July 17 2014.

[7] " ns-3 " , URL: https://www.nsnam.org/ [accessed: 2017-8-18].

Evaluation of Safety and Efficiency Simulation of Cooperative Automated Driving through Intersection

Kenta Kimura
Computer and Information Science
Graduate School of
Science and Engineering
Doshisha University
Kyoto, Japan
Email:kenta.kimura@nislab.doshisha.ac.jp

Shuntaro Azuma
Computer and Information Science
Graduate School of
Science and Engineering
Doshisha University
Kyoto, Japan
email:syuntaro.azuma@nislab.doshisha.ac.jp

Kenya Sato
Computer and Information Science
Graduate School of
Science and Engineering
Doshisha University
Kyoto, Japan
email:ksato@mail.doshisha.ac.jp

Abstract—**The research and development of automated driving are recently thriving, and a mechanism continues to progress through which the car itself assumes the role of the driver. However, when automated driving only uses information that is collected by onboard sensors, no information on vehicles in the intersecting lanes can be obtained from intersections where the visibility is often bad and suffers from large potential blind spots. This situation increases the risk of collisions. Since no information can be obtained on blind spots until entering the intersection, the vehicle must temporarily slow down to confirm the safety of the situation. To improve safety and efficiency, information must be obtained for each vehicle and communicated with the surrounding vehicles. In this research, we examine safety and efficiency by comparing cases of automated vehicles with and without communication when entering an intersection with poor visibility. We evaluated with a simulator and identified safety and efficiency effects when an automated vehicle uses communication at an intersection that features poor visibility.**

Keywords–cooperative automated driving; V2V communication; traffic flow.

I. INTRODUCTION

The research and development of automated driving have increased in recent years. A camera, laser radar, and milliwave radar are mounted on an autonomous automated vehicle for collecting peripheral information. Then the vehicle's operation is controlled using the surrounding environment information. However, such in-vehicle sensors have drawbacks because detection is impossible outside the range of the viewing angles, and so avoiding collisions is difficult at intersections that suffer from poor visibility.

With Vehicle-to-Vehicle (V2V) communication, blind spot information can be acquired that the vehicle cannot see directly. To operate safely using this information, research on cooperative automated driving is being conducted. The recognition rate near 300 m increases by sharing the host vehicle's information and the sensor information using V2V communication instead of automated driving that relies solely on sensor information [1].

The level of automated driving techniques has already been defined by Society of Automotive Engineers (SAE) international. Level 2 vehicles, in which the automated vehicle partially controls the vehicle's operation, are beginning to appear on the market. For example, when a vehicle predicts an accident, it automatically brakes. This is not completely automated driving; it just illustrates the scope of driving support. In this research, we evaluate automated vehicles that can operate such vehicles whose popularity is expected to increase in the future. In this research, we determine the safety criterion for passing through intersections when using communication and compare cases with and without sharing the surrounding information of a vehicle. Based on our results, we evaluate the influence of shared communication on traffic efficiency and safety when passing through an intersection that suffers from poor visibility.

The remainder of this paper is structured as follows. Section II details of problems at intersections with blind spots. Section III details of the calculation method when vehicles pass through the intersection. Section IV details the evaluation of the proposal. Section V details the consideration obtained from the evaluation results. Section VI details the conclusion.

II. PROBLEMS AT INTERSECTIONS WITH BLIND SPOTS

Accidents at intersections with poor visibility are a problem. According to official police statistics of traffic accident occurrences, urban intersections are the most common place for such accidents [2]. The number of accidents occurring at Japanese urban intersections in 2016 is 208,404. Since the number of accidents occurring at non-urban intersections is 46,952, there are many accidents at urban intersections. As shown in Figure 1, at intersections with poor visibility, the blind spots are large. At such intersections, since the probability of crossing collisions is high, drivers must pause before entering intersections and pass through them only after confirming that they are safe. This action is necessary regardless of the presence or absence of a vehicle in the intersecting lane. If there is no vehicle in the intersecting lane, the time required to confirm safety becomes wasted as the vehicle passes through the intersection. These situations and decisions are identical for automated vehicles.

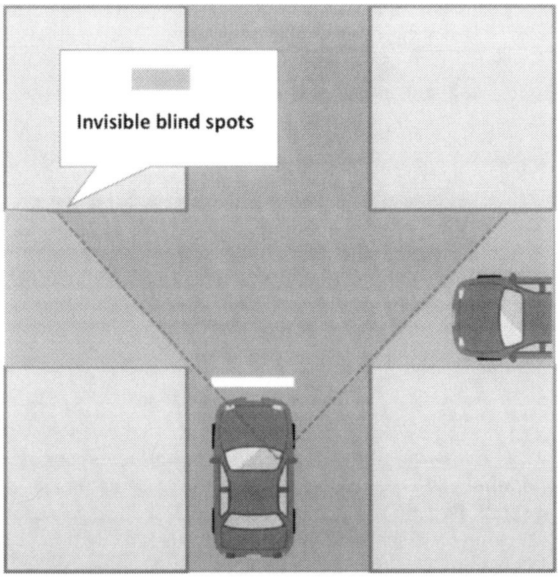

Figure 1. Intersection blind spot

III. PROPOSAL

A. Outline

Whether vehicles passing through intersections will collide with vehicles traveling from intersecting lanes must be verified. When no communication is used and the vehicle enters an intersection without traffic lights, it pauses before entering it and checks whether it might collide with a vehicle in the intersecting lane. If no collision is imminent, it passes through the intersection. At an intersection with traffic lights, vehicles run based on the signals. In this research, we evaluate passing through intersections based on the premise that accurate position information and speed information can be acquired using V2V communication.

B. Precondition

We set the following preconditions:

1) Automated vehicles can communicate with each other.
2) Communication loss is ignored.
3) Position information error is ignored.
4) Communication is shared every 0.1 seconds.
5) The crossing lanes are blind spots and invisible.

C. TTC

As a criterion for passing through an intersection, we use Time-To-Collision (TTC) [3]. As shown in Figure 2, the position and speed of the following and preceding vehicles are defined as x_f, v_f, x_p, and v_p. If TTC is defined as t_c, it can be expressed by (1):

$$t_c = -\frac{x_f - x_p}{v_f - x_p} \quad (1)$$

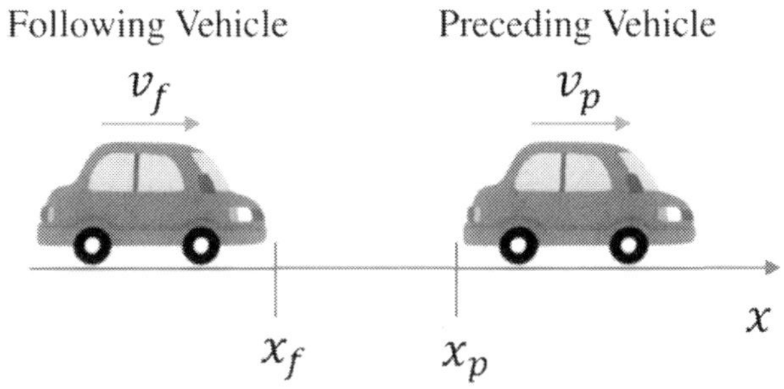

Figure 2. TTC Value

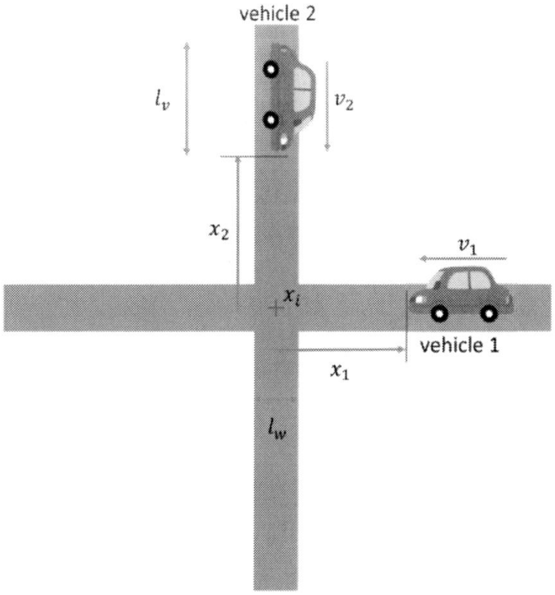

Figure 3. Value at intersection

D. TTC in This Research

However, since TTC is an index for vehicles in the same lane, the formula must be converted into a calculation between vehicles on crossing lanes. Each value at the intersection is shown in Figure 3. Here, a vehicle on a non-priority road is defined as vehicle 1, and a vehicle on a priority road is defined as vehicle 2. We also define the distance to the center of the intersection as x_1 and x_2, the speed as v_1 and v_2, the length of the vehicle as l_v, and the length within the intersection as l_w.

First, we calculate the time until vehicle 1 reaches the intersection's entrances. If this time is assumed to be t1, it can be obtained by (2). The distance to the intersection's entrance can be obtained by subtracting half of the intersection's width from its center distance:

$$t_1 = \frac{x_1 - \frac{l_w}{2}}{v_1} \quad (2)$$

Next, we calculated the time it takes for vehicle 1 to completely pass through the intersection. If this time is defined as t_2, it can be obtained by (3), and t_2 is the time obtained by adding t_1 to the time required for vehicle 1 to travel the total distance of the width of the roadway and the length of the vehicle:

$$t_2 = t_1 + \frac{l_w - l_v}{v_1} \quad (3)$$

Next, we calculate the position of vehicle 2 of time t_1 and t_2. Here the position during each time lapse is defined as $x_{2,1}, x_{2,2}$, obtained by (4) and (5):

$$x_{2,1} = x_2 + v_2 t_1 \quad (4)$$

$$x_{2,2} = x_2 + v_2 t_2 \quad (5)$$

The TTC value that is allowed when crossing an intersection is defined as t_{ttc}, which determines whether the position of vehicle 2 is dangerous when vehicle 1 passes through the intersection. x_i is the center position of the intersection. When either (6) or (7) is satisfied, it is dangerous for vehicle 1 to pass through the intersection, and passage is denied:

$$x_i - \frac{l_w}{2} - v_2 t_{ttc} \leq x_{2,1} \leq x_i + \frac{l_w}{2} + v_2 t_{ttc} + l_v \quad (6)$$

$$x_i - \frac{l_w}{2} - v_2 t_{ttc} \leq x_{2,2} \leq x_i + \frac{l_w}{2} + v_2 t_{ttc} + l_v \quad (7)$$

Vehicles that are not allowed to cross the intersection will be stopped before they enter it.

IV. EVALUATION

A. Simulator

Our evaluation uses Vissim [4], a microscopic multi-modal traffic flow simulation software package developed by Planung Transport Verkehr (PTV) AG in Karlsruhe, Germany, that can extract such problems as congestion and the influence of road construction. It can also visually confirm a set simulation with 3D graphics.

Vissim also supports the Component Object Model (COM) interface through which it can communicate with external applications and scripts. Using this function, we can set input data to Vissim and obtain output data from it. In this research, we collected vehicle information from Vissim using Python 2 and calculated TTC.

B. When Only a Straight Traveling Vehicle is being Operated

In this section, we evaluate the case where the vehicle does not make a left or a right turn and only moves forward toward an intersection.

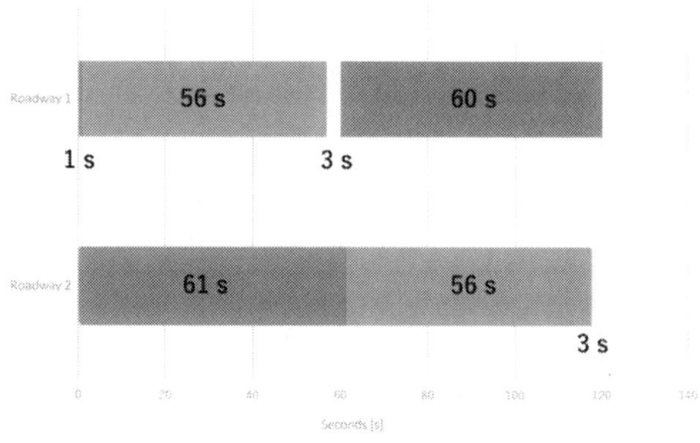

Figure 4. Traffic light setting

1) Evaluation method: According to technical government guidelines [5], the criterion for braking in automated braking systems is a TTC of 1.4 seconds or less for passenger cars. Therefore, we set the TTC value of this intersection's passing criterion to 1.4 seconds and calculated the intersection passing determination from the timing when the vehicle enters within around 100 m of the intersection.

Table I details the setting. The intersection has one lane on each side. One of the lanes in it is the priority road, and the vehicle on the non-priority road determines the passing through the intersection based on the TTC. The speed limit is 50 km/h.

We simulated two other models to measure the effect of passing through the intersection using V2V communication.

The first model does not communicate. The vehicle on the non-priority road side pauses for 0.5 seconds to confirm the intersection's safety before entering it. If it is safe, then it passes through the intersection. The safety criterion is judged by whether the vehicle from the priority road side is approaching within 100 m from the intersection.

In the second model, the vehicle does not communicate and advances based on the intersection's traffic light. The traffic light's cycle is shown in Figure 4. One cycle is set to 120 seconds: 2 minutes for the signal's total time, 1 seconds for red, 56 seconds for green, and 3 seconds for yellow.

The third model measures a vehicle's travel time on a 1000-m non-priority road that includes an intersection. Travel time refers to the time spent driving on a specified section.

2) Evaluation results: We conducted three different types of evaluations and measured the travel times of the three models in the above intersection.

In the first evaluation, when the number of vehicles in each lane was set to 500 vehicles/hour, we measured the travel

TABLE I. SIMULATION PARAMETERS

Simulator	Vissim 9.00-09
Number of vehicles in one lane per hour	100~700 vehicles per hour
Measurement time	10 minutes
Measurement section	1000 m
Road width	7 m
Center position of intersection	500 m position
Speed limit	50 km/h

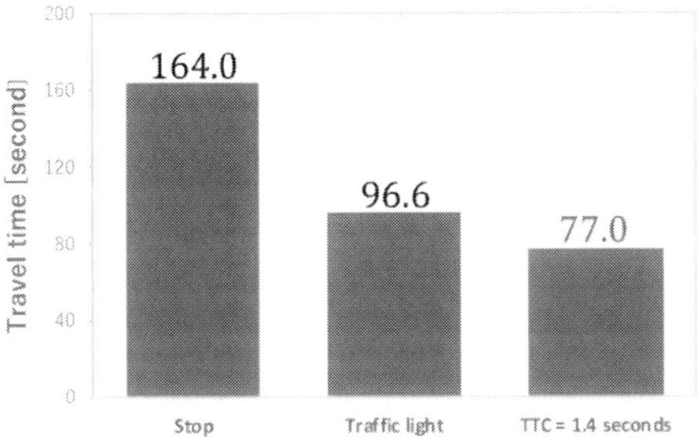

Figure 5. Travel time by model

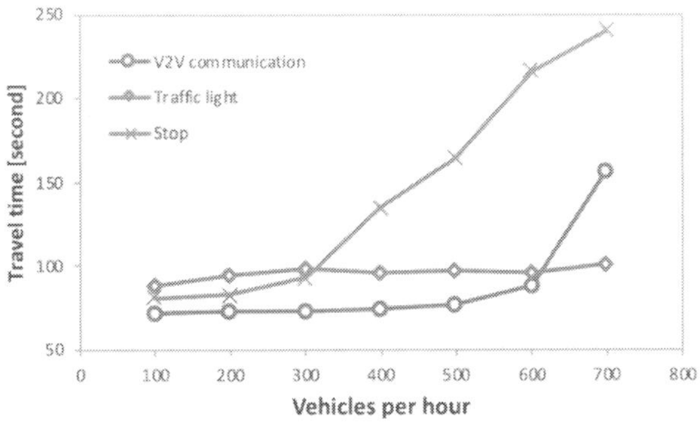

Figure 6. Travel time by number of vehicles

time required for a vehicle to travel 1000 meters on the non-priority road side. As described above, 1.4 seconds is the safety criterion used by TTC for calculating safe passage through the intersection using communication (Figure 5). The traveling time was short in the following order: the model using the proposed method using communication, the model using the traffic light, and the model based on the determination of safety by pausing.

In the second evaluation, the number of vehicles per hour in each lane was increased in increments of 100. The result is shown in Figure 6. The travel time when controlled by signals did not significantly affect the travel time in the number of vehicles in the measured range. However, in the models that did not use both communication and traffic signals, the traveling time increased as the number of vehicles increased. When using communication, the travel time did not change greatly from 100 vehicles per hour to 600 vehicles per hour. But if the number of vehicles per hour increased to 700, the travel time increased significantly.

For the third evaluation, we measured the travel time every 0.2 seconds from 1.4 to 2.0 seconds for the TTC seconds used for judging passage through the intersection using V2V communication. For passenger vehicles and larger vehicles, the distance from braking to actually stopping is different. In the case of larger vehicles, operating the automated brake system is desirable when the TTC is 1.6 seconds or less [6]. Therefore, we measured the travel time by changing the TTC value (Figure 7). The higher the TTC value is, the higher is

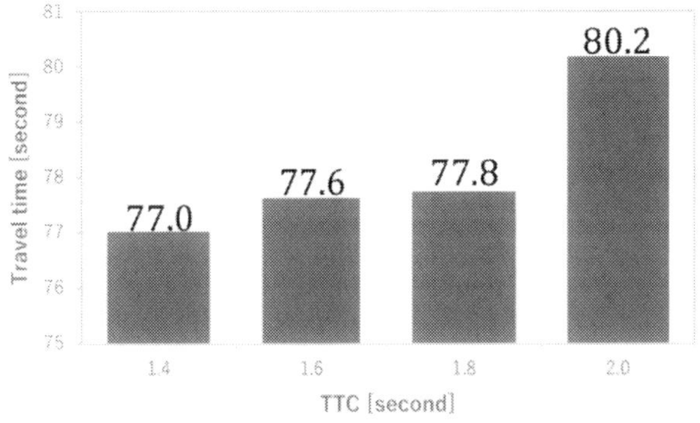

Figure 7. Travel time by TTC

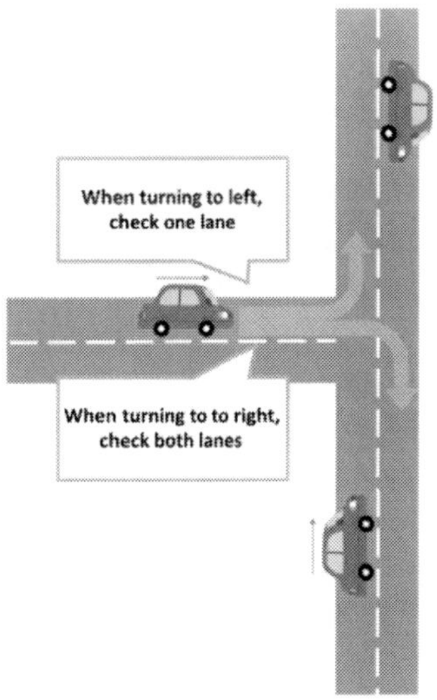

Figure 8. Shape of T-junction

the travel time value.

C. At a T-junction

In this subsection, we evaluated when a vehicle on a non-priority road makes a left or a right turn at a T-junction. As a precondition, the vehicles on the priority road side must go straight ahead. Vehicles are driven on the left - hand side of the road.

1) Evaluation method: The T-junction ' s shape is shown in Figure 8. We can judge whether a left turn is possible by checking the safety of one lane when turning left. However, when turning right, the safety of both lanes must be checked, and the conditions for passing through the intersection become stricter than when turning left.

For the evaluation, we compared the proposed method ' s model, a model that paused at the intersection, and a model that obeyed the traffic light. Pause and signal period settings were evaluated with settings that resemble those in the previous section. The simulation settings are shown in Table II. We established an intersection 500 m from the point of the vehicle and measured the travel time when the vehicle on the non-priority road that makes a left or a right turn travels 1000 m. Here the vehicle traveling on the non-priority road chooses left

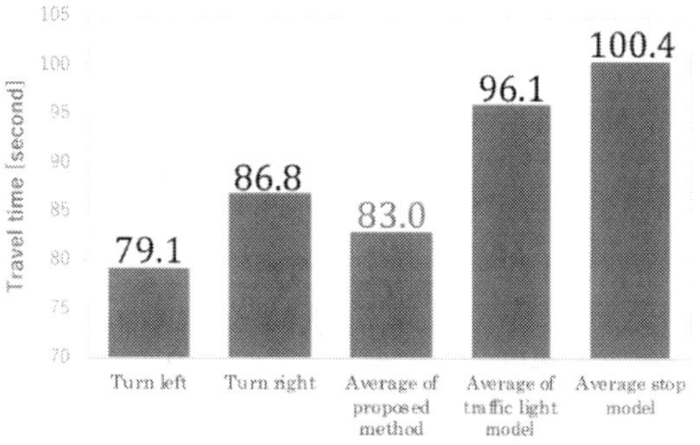

Figure 9. Travel time of T-junction

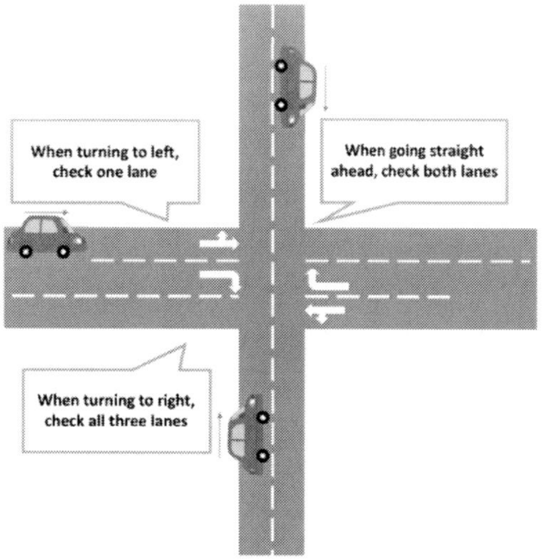

Figure 10. Shape of crossroad

or right turns with a 50 % probability. Then we measured the travel time by the traveling direction.

2) Evaluation result: The evaluation result is shown in Figure 9. The time to travel 1000 m for left turns is 79.1 seconds and the travel time for right turns is 86.8 seconds. This result suggests that the safety criterion is more severe and the travel time is longer when turning right than left.

D. At a Crossroad

Next, we evaluate when the vehicle on the non-priority road side makes a left or a right turn or continues straight ahead and travels on the crossroad. As a precondition, the vehicles on the priority road side should go straight ahead. Vehicles are again driven on the left - hand side of the road.

TABLE II. SIMULATION PARAMETERS FOR T-JUNCTION AND CROSSROAD

Number of vehicles in one lane per hour	300 vehicles per hour
Measurement time	10 minutes
Measurement section	1000 m
Road width	7 m
Center position of intersection	500 m position
Speed limit	50 km/h
TTC value of intersection passing criterion	1.4 seconds

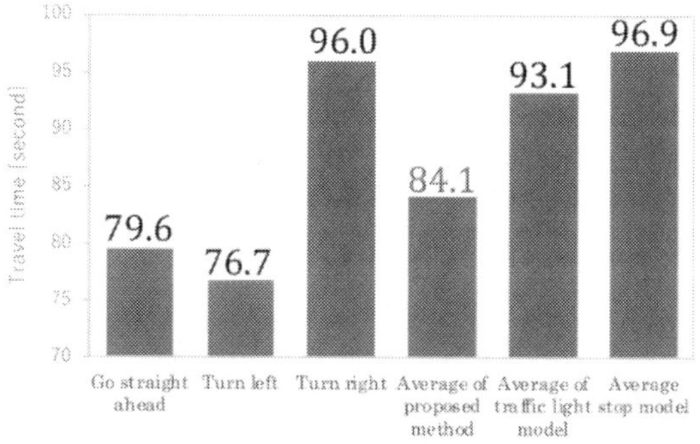

Figure 11. Travel time of crossroad

1) Evaluation method: The shape of the crossroads is shown in Figure 10. It is necessary to check the safety of one lane when turning left, two lanes when going straight, and three lanes when turning right. Thus, the safety criterion depends on the traveling direction, and the criteria become stricter in the order of left turn, straight run, and right turn. When turning right, the traveling direction of the oncoming vehicle must be confirmed. However, when right-turn vehicles face each other, no collision occurs in the intersection. Therefore, in addition to the vehicle ' s speed and position information, information of the traveling direction must be shared. Unlike the other evaluations, the condition is set to share information of traveling directions.

For our evaluation, we compared the following models: our proposed method, one that paused before entering the intersection, and one with traffic lights. The pause settings and the signal period settings are evaluated with settings that resemble those used in the previous section. We established an intersection 500 m from the vehicle and measured the travel time at 1000 m. We assume that a vehicle on the non-priority road side is selected with a probability of 70 % for straight ahead, 20 % for a left turn, 10 % for a right turn, and measured the travel time for each traveling direction.

2) Evaluation result: The evaluation result is shown in Figure 11. The traveling time increased as the number of lanes whose safety must be confirmed also increased in the crossroad.

V. CONSIDERATION

A. When Only a Straight Traveling Vehicle is being Operated

From Figure 5, using V2V communication at an intersection where only straight-ahead vehicles run reduces the travel time. Traffic efficiency is defined as the time to reach a destination. With V2V communication, travel time was reduced and efficiency was improved. TTC calculation verified the collision delay time with the vehicle in the lane that intersects when passing through the intersection and improved safety more than without communication.

From Figure 6, the travel time of the models using V2V communication is the shortest when the number of vehicles per hour ranged from 100 to 600. However, the travel time is longer than in the model with traffic signals if the number of generated vehicles increased to 700. A vehicle on a non-priority road cannot cross the intersection because it runs on the priority road without interruption. If a certain traffic volume is exceeded, negotiation is required at the intersection. For example, one method is to yield at an intersection to a vehicle that has stopped for a certain period before entering an intersection.

According to the result in Figure 7, the travel time also increases as the TTC value is increased. Although the risk of collisions is reduced by widening the distance (that remains safe) between vehicles, the travel time's efficiency is degraded. Therefore, safety and efficiency have a trade-off relationship. The TTC needs to set a value that maximizes the efficiency by ensuring a minimum level of safety.

B. For a T-junction

Considering the left or right turn of a vehicle on a non-priority road, the travel time at a T-junction is shown in Figure 9. The traveling time for a right turn is about seven seconds longer than for a left turn. The average travel time of left or right turns is 83.0 seconds. The vehicle's average speed in the 1000-m section is about 43.4 km/h, which is relatively fast since the speed limit is 50 km/h. Our proposed method is effective for traffic efficiency at T-junctions.

C. In Case of Crossroad

Figure 11 shows that the traveling times increase as the number of lanes whose safety must be confirmed increases. The average travel time of the proposed method is 84.1 seconds, meaning that the average speed of a vehicle in the 1000-m section is about 42.8 km/h. Since the speed limit is 50 km/h, this level is fast. In addition, the model's travel time using traffic signals is 93.1 seconds, and model's travel time that pauses is 96.9 seconds. Since the travel time of the proposed method is the shortest, it is effective for the traffic efficiency of crossroads.

VI. CONCLUSION

In recent years, research on automated driving has been increasing. The information that can be obtained by an automated vehicle as a single unit is limited, and it is impossible to collect information on blind spots when viewed from the vehicle. We must supplement the missing information by sharing information with various objects. We verified the safety and efficiency when using cooperative automated vehicles at an intersection where the outlook is bad and blind spots are large. With communication, the speed and position information of vehicles around the intersection are acquired to judge whether safe passage is possible.

Evaluation results showed that efficiency and safety were improved more than the case of confirming the safety of passing through the intersection without communication when it confirms safe passage.

ACKNOWLEDGMENT

This work was supported in part by JSPS KAKENHI Grant Number 16H02814.

REFERENCES

[1] H. Gunther, R. Riebl, L. Wolf, and C. Facchi, "Collective Perception and Decentralized Congestion Control in Vehicular Ad-hoc Networks", Vehicular Networking Conference(VNC), 2016.

[2] National Police Agency, "Occurrence of traffic accident accident in 2017", URL: https://www.npa.go.jp/toukei/koutuu48/toukei.htm[accessed : 2018-04].

[3] T. Hiraoka and S. Tanaka, "Collision Risk Evaluation Based on Deceleration for Collision Avoidance", The Society of Instrument abd Control Engineers, Vol.47, No.11, 2011, pp.534-540.

[4] "PTV Vissim", URL: http://vision-traffic.ptvgroup.com/en-us/products/ptv-vissim/[accessed : 2018-04].

[5] Ministry of Land, Infrastructure, Transport and Tourism, "Reference material - Ministry of Land, Infrastructure, Transport and Tourism", URL: www.mlit.go.jp/common/000162826.pdf[accessed : 2017-8].

[6] T. Sawada, T. Hirose, N. Kasuga, and M. Zeniya, "Study on Evaluation of Collision Mitigation Brake System", International Association of Traffic and Safety Sciences, Vol.33, No.4, 2008, pp.311-338.

Evaluation of a Method for Improving Pedestrian Positioning Accuracy using Vehicle RSSI

Yuya Nishimaki
Computer and Information Science
Graduate School of Science and Engineering
Doshisha University
Kyoto, Japan
email:yuya.nishimaki@nislab.doshisha.ac.jp

Hisato Iwai
Electrical and Electronic Engineering
Graduate School of Science and Engineering
Doshisha University
Kyoto, Japan
email:iwai@mail.doshisha.ac.jp

Kenya Sato
Computer and Information Science
Graduate School of Science and Engineering
Doshisha University
Kyoto, Japan
email:ksato@mail.doshisha.ac.jp

***Abstract*—As the penetration rate of smartphones and tablet-type devices increases, various services using such location information are being used. In navigation applications, we can check our current location and how to get to a destination. Even if we are in a relatively new area, we can go anywhere using a navigation application. Other systems prevent traffic accidents by exchanging position information using vehicle-to-pedestrian communication. In these services and systems, especially for preventing accidents, accurate position information is critical. Currently, the Global Positioning System (GPS) is most frequently used as a positioning method to acquire position information outdoors, but its signals are influenced by the surrounding buildings in urban areas, reducing positioning accuracy. Therefore, in this paper, we propose a position estimation method using the Received Signal Strength Indication (RSSI) of vehicles and beacons for high and stable positioning accuracy outdoors. We applied a Kalman filter to RSSI and dynamically calculated the path loss index using vehicle-to-vehicle communication. We compared the positioning accuracy of our method and conventional methods by simulations and showed our method's superiority.**

Keywords–position estimation; vehicle-to-vehicle communication; RSSI; path loss index.

I. INTRODUCTION

As the penetration rate of smartphones and tablet-type devices continues to increase, various services using their location information are being used. For example, we can check our current location on a map, look up a route from it to a destination, and get surrounding shop information and coupons. However, if the positioning error is too large, we might get lost or fail to get the information we want. In recent years, Intelligent Transportation Systems (ITS) have also been investigated that improve traffic safety, efficiency and driving comfort. For example, through vehicle-to-vehicle and vehicle-to-pedestrian communication, related work makes efforts to prevent vehicles from colliding with vehicles and pedestrians through exchange of information such as position and speed [1][2]. Since the large positioning error might lead to accidents, we need accurate position information of every involved pedestrians and vehicle to reduce traffic accidents and maintain safety. Among positioning systems that acquire position information which is important in such services and systems, Global Positioning System (GPS) is most frequently used. Its positioning accuracy ranges from several meters to several tens of meters. But in urban areas that are littered with high-rise buildings, GPS signals are blocked by buildings and influenced by multipaths, further increasing the positioning error [3]. Since GPS accuracy is affected by the surrounding environment, achieving stable positioning is difficult.

In this research, we focus on a pedestrian whose positioning error is often larger compared to vehicles. We propose a method that uses vehicles to improve the outdoor positioning accuracy of pedestrians, since both vehicle-to-vehicle and vehicle-to-pedestrian communication will become more widespread due to ITS development. The rest of this paper is organized as follows. In Section II, we show conventional position estimation methods and their problems. In Section III, the proposed method is described. Simulation and evaluation results are presented in Section IV. Finally, Section V gives the conclusion.

II. CONVENTIONAL POSITION ESTIMATION METHOD AND PROBLEMS

A. GPS

In a GPS, which is the most widely used positioning method. A device receives signals with time information transmitted from multiple GPS satellites and estimates a position using pseudo distances obtained from the differences between transmission and reception times. In line-of-sight places and areas without high buildings around the target, since it can receive signals from many GPS satellites, GPS can estimate a position with error within a few meters. In non-line-of-sight places and urban areas with many high-rise buildings, GPS signals are influenced by shielding, reflection, and diffraction by obstacles. Sometimes positioning error becomes several meters to several tens of meters.

B. RSSI-based position estimation

The positions of smartphones and tablet-type devices can be estimated by RSSI when receiving radio waves transmitted from such beacons as Bluetooth Low Energy (BLE) or WiFi [4][5]. As shown in Figure 1, when the target acquires the RSSI of Beacons 1 to 4, first, we calculate the distance between the target and each beacon from the RSSI and find the target's position by triangulation using the distance and the position of each beacon. We can calculate the distance from the RSSI

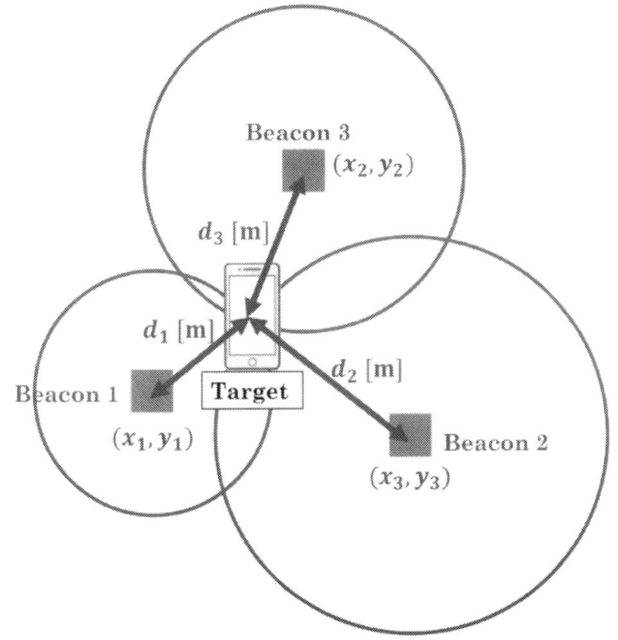

Figure 1. RSSI-based positioning

by a property through which RSSI is attenuated as the distance increases. In general, we can model RSSI's attenuation (as in (1)) to find the distance between the target and each node:

$$P(d) = A - 10n\log_{10} d. \quad (1)$$

Here, $P(d)$ is the RSSI [dBm] at position d [m] away from the transmission source, A is the RSSI [dBm] 1 m from the transmission source, and n is the path loss index. The path loss index is a value that represents the degree of attenuation of the radio waves based on the distance, and in the free space, $n = 2$. Here, RSSI attenuates in proportion to the square of the distance.

C. *Problems when using RSSI*

For an accurate position estimation using RSSI, the distance to each node must be accurately measured. However, since it is sometimes impossible to obtain the ideal RSSI in an actual environment, the distance calculated from the RSSI includes error. There are several causes for the error increase. One is that the path loss index is constantly changing due to the surrounding environment of the radio wave propagation. Another is that RSSI fluctuates. Although n in (1) is theoretically a constant value, it should be set dynamically because it is always changing, too. Also, RSSI does not always become a constant value even if the distance between the sender and the receiver is invariant. It varies with time due to the following factors:

- Fading

Fading, which is the RSSI fluctuation received by wireless communication, occurs during communication while moving by such mobile communication as mobile devices. In environments with many scattered materials, the radio waves transmitted from the sender are reflected, diffracted, and scattered by buildings and moving objects, causing them to arrive at the receiver with a time difference. Interference of these radio waves also causes fading.

- Shadowing

This RSSI fluctuation, which occurs when a shielding object exists between a sender and a receiver, follows a lognormal distribution.

- Co-channel interference

This RSSI fluctuation occurs when a device using the same frequency band operates in the vicinity.

For accurate estimation of distance, it is necessary to reduce the influence of RSSI fluctuation and to use the dynamic path loss index.

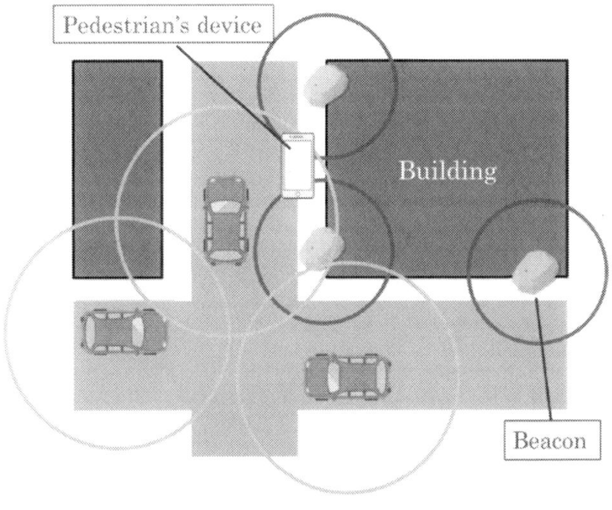

Figure 2. Outline

III. PROPOSAL

A. *Outline*

In urban areas, positioning error might increase when GPS signals are blocked by buildings and influenced by multipaths. In this research, we propose a method to improve the positioning accuracy of outdoor pedestrians and outline it in Figure 2. We estimate a pedestrian's position using the RSSIs of vehicles and beacons on the road side. When we assume that the vehicle regularly transmits its own information to its surroundings, pedestrians can receive radio waves of vehicles in addition to beacons, allowing them to also use RSSIs from their vehicles to estimate their positions. However, the RSSIs of beacons and vehicles suffer from the problems shown in Section II-C. To solve them, we apply filtering to RSSIs and dynamically calculate the path loss index using vehicle-to-vehicle communication.

B. *Presuppositions*

1) All of the vehicle position information is accurate.
2) The pedestrian's device can receive 700-MHz band radio waves.
3) The vehicle regularly transmits its own information to the surroundings.

TABLE I. PARAMETERS FOR VEHICLES

Pedestrian state	$\hat{P}(0)$	Q	R
Stationary	1000	4.44	25.44
Moving	1000	5.37	27

TABLE II. PARAMETERS FOR BEACONS

Pedestrian state	$\hat{P}(0)$	Q	R
Stationary	1000	0.00046	19.0454
Moving	1000	5.41	13.3

C. RSSI filteration with Kalman filter

We used a Kalman filter as a filtering method to reduce the influence of fading, among variation factors mentioned in Section II-C. We applied it to the RSSI received by pedestrians from both the beacon and the vehicle, and did filtering using the following formulas. In the Kalman filter in the prediction step, we obtained RSSI's prior state estimate at the current time using the information of one time before. In the filtering step, we modified RSSI's prior state estimate by an observation value to find the posteriori state estimate:

- Prediction step

$$\hat{x}^-(k) = \hat{x}(k-1) \quad (2)$$
$$\hat{P}^-(k) = \hat{P}(k-1) + Q \quad (3)$$

- Filtering step

$$g(k) = \frac{\hat{P}^-(k)}{\hat{P}^-(k) + R} \quad (4)$$
$$\hat{x}(k) = \hat{x}^-(k) + g(k)(y(k) - \hat{x}^-(k)) \quad (5)$$
$$\hat{P}(k) = (1 - g(k))\hat{P}^-(k). \quad (6)$$

Here, $\hat{x}^-(k)$ and $\hat{x}(k)$ are prior and posteriori state estimates. $\hat{P}^-(k)$ and $\hat{P}(k)$ are prior and posteriori error variances, Q is a system noise variance, R is a measurement noise variance, $y(k)$ is a measurement value, and $g(k)$ is the Kalman gain. In this study, we set $\hat{P}(0)$, Q, and R to the values shown in Tables I and II. We determined them by simulation experiments in advance.

D. Calculation of dynamic path loss index

$$n_i = \frac{A - P(d_i)}{10 \log_{10} d_i} \quad (7)$$

$$n_k = \frac{\sum_{i=1}^{m} n_i}{m} \quad (i \neq k). \quad (8)$$

When vehicle k is communicating with m neighboring vehicles in vehicle-to-vehicle communication, the path loss index between vehicles k and i $(1 \leq i \leq m)$ can be calculated by (7), which is obtained by transforming (1). Here, d_i represents the distance [m] between vehicles k and i, and $P(d_i)$ represents RSSI [dBm] from vehicles i. We assume that the vehicle-to-vehicle communication is within the line-of-sight and use RSSI exceeding $P(d_i) > -50\,\mathrm{dBm}$. Then with (8), we calculated the average value of the path loss indices for each surrounding vehicle within a certain time and use the result as the path loss index around vehicle k. The procedure for dynamically calculating the path loss is described below, and Figure 3 describes the calculation example.

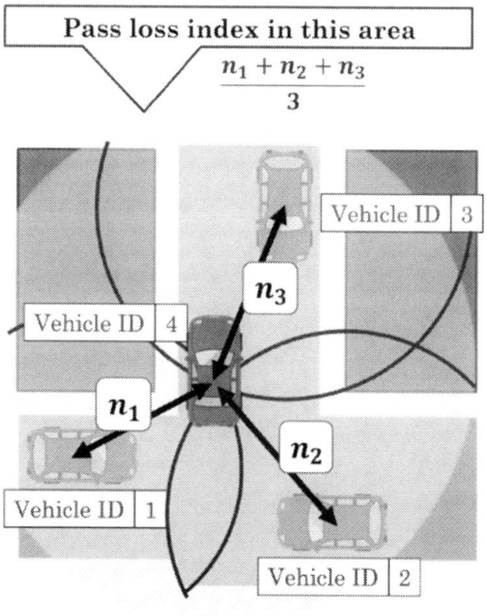

Figure 3. Example of calculation of dynamic path loss index

1) When a vehicle receives packets from a surrounding vehicle, it obtains the distance to the sender using the sender's position and its own position.
2) The vehicle obtains the path loss index between the sender and itself using (7) and stores the result and the sender's vehicle information in the vehicle information management table.
3) At the time of transmission, the vehicle calculates the average path loss index within the past $500\,\mathrm{ms}$ in the vehicle information management table by (8) and regards the average as the path loss index around itself.

E. Positioning algorithm

Next, we describe the algorithm that calculates the pedestrian's position using RSSI acquired from the beacons and the vehicles. If the pedestrian gets information of m $(m \geq 1)$ nodes of the beacons and the vehicles, she can calculate the distance with each node from the RSSI by (9), which is obtained by transforming (1).

$$d_i = 10^{\frac{A - P_i}{10n}} \quad (1 \leq i \leq m). \quad (9)$$

The smaller an RSSI is, the larger is the variation and the error from the theoretical values [6]. Therefore, instead of using all the acquired RSSIs for position estimation, we set thresholds and in advance excluded those with large error from the theoretical value. We set the threshold and the parameters for distance calculation (Table III).

We also selected the RSSI depending on the state of the pedestrian and the vehicle.

- When pedestrian and vehicle are stationary

TABLE III. THRESHOLD AND PARAMETERS

Node type	*Threshold*	*A*	*n (Path loss index)*
Beacon	−81 dBm	−61 dBm	2.0
Vehicle	−50 dBm	−10.816 dBm	dynamic value

When determining the distance to the beacon or the vehicle, we used the maximum RSSI above the threshold received within the past 1 s. This is because the maximum RSSI is less influenced by fading than the others, and the distance error also becomes smaller.

- When pedestrian is stationary and vehicle is moving

When obtaining the distance to the beacon, the process is the same as when pedestrian and vehicle are stationary. When calculating the distance to the vehicle, we use the latest RSSI among those received after the final position's estimation time.

- When pedestrian is moving

When calculating the distance to the beacon or the vehicle, we use the latest RSSI among those received after the final position estimation time.

We determined the pedestrian position using Weighted Centroid Localization (WCL) [6]. (10) to (12) show how to calculate the position by WCL. (x_i, y_i) is the position of the i th node that corresponds to the selected RSSI, and w_i is the weight, and we obtained the weighted average of the position of each node. The weight is the reciprocal of the distance obtained from RSSI, and the g value is 1.5 to minimize the positioning error in (12):

$$x_w = \frac{\sum_{i=1}^{m} x_i w_i}{\sum_{i=1}^{m} w_i} \tag{10}$$

$$y_w = \frac{\sum_{i=1}^{m} y_i w_i}{\sum_{i=1}^{m} w_i} \tag{11}$$

$$w_i = \frac{1}{d_i^g}. \tag{12}$$

IV. EVALUATION AND CONSIDERATION

A. Simulator

We used Scenargie as a simulator to evaluate our proposed method's performance. Scenargie is a network simulator developed by Space-Time Engineering (STE) [7]. By combining expansion modules, various simulations like LTE, vehicle-to-vehicle communication and a multi-agent simulation can be constructed. Since communication systems and evaluation scenarios are becoming more complicated, this ingenious simulation greatly reduces the effort required to create them. Examples include a GUI scenario creation, map data, the graphical information display of a communication system, and a radio wave propagation analysis function.

B. Evaluation Model

Since this research's goal is improving the positioning accuracy of pedestrians in urban areas, we did our simulation in an evaluation environment where pedestrians are surrounded by buildings. The simulation parameters are shown in Table IV, and its environment is shown in Figure 4. GIS-BASED-RANDOM-WAYPOINT in Table IV is a model where each vehicle randomly determines a passing point, moves along the road, and passes through it. The ITU-R P.1411 model [8] is a radio wave propagation scheme that considers road map information. Since radio waves are attenuated based on the road's shape, this model closely resembles reality compared with a two-ray model using direct waves and reflected waves from the ground.

TABLE IV. SIMULATION PARAMETERS

Simulator	Scenargie2.1	
Simulation time	20 [s]	
Beacon intervals	5, 10, 15, 20 [m]	
Number of vehicles	0, 10, 20, 30, 40, 50, 60, 70, 80	
Vehicle mobility model	GIS-BASED-RANDOM-WAYPOINT	
Vehicle velocity	15 ∼ 20 [m/s]	
Pedestrian velocity	Stationary, 2 [m/s]	
	Beacon	Vehicle
Transmission power	-21 [dBm]	19.2 [dBm]
Frequency bands	2.4 [GHz]	760 [MHz]
Communication intervals	500 [ms]	100 [ms]
Propagation model	TwoRayGround	ITU-R P.1411
Fading model	Rayleigh	

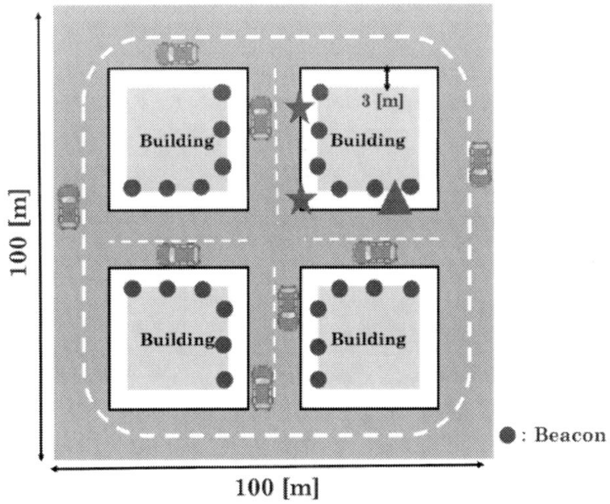

Figure 4. Evaluation environment

C. Evaluation items

- Comparison of positioning error with conventional methods

Here, we label GPS as Conventional 1 and the method that only uses beacons as Conventional 2. Proposal 1 is our proposed method, and Proposal 2 does not adopt filtering and uses a static path loss index. In the simulator, since we cannot measure the GPS-positioning accuracy, we compared 15 m as GPS positioning error [3].

- Evaluation of beacon intervals

We evaluated the positioning error when the beacon interval increases in 5-m increments from 5 to 20 m. At this time, the number of vehicles was 50.

- Evaluation of number of vehicles

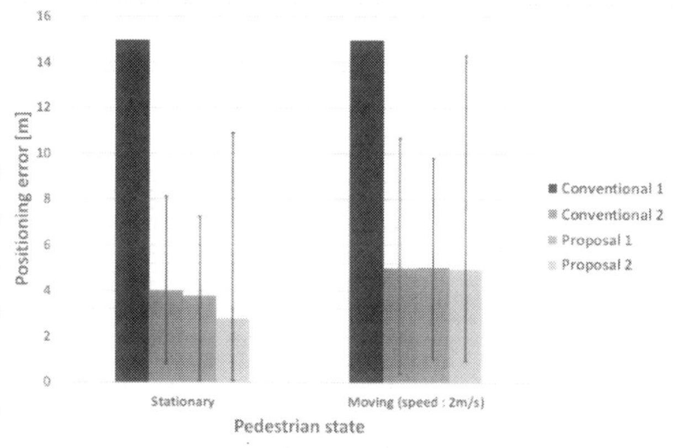

Figure 5. Average, maximum and minimum positioning error in conventional methods and proposed methods

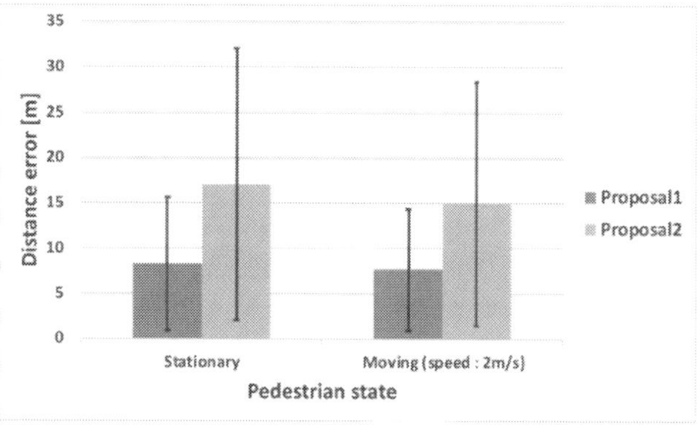

Figure 6. Distance error and standard deviation by pedestrian state of Proposal 1 and Proposal 2

We evaluated the positioning error when the number of vehicles increased from 0 to 80 in increments of 10. In this case, the beacon interval was $10\,\mathrm{m}$.

Regarding the positioning error, when a pedestrian is stationary, we simulated at 5-m intervals between the stars shown in Figure 4 and calculated their average positioning error. We also simulated in a situation where the pedestrian is moving at a speed of $2\,\mathrm{m/s}$ from the top stars to the triangle in Figure 4 and calculated the average positioning error.

D. Comparison of positioning error with conventional methods

Figure 5 shows the positioning error of the pedestrian in the conventional and proposed methods. With the proposed method, the average positioning error decreased more for both stationary and moving pedestrians than in Conventional 1. Compared to Conventional 2, in Proposal 1 the positioning error in the moving situation hardly changed, but the average positioning error in the stationary situation was smaller. Moreover, the maximum positioning error became smaller both in the stationary and moving situations. Compared to Proposals 1 and 2, the average positioning error in Proposal 2 was smaller than in Proposal 1. However, the variation in the positioning error and the maximum positioning error in both the stationary and moving situations in Proposal 1 was smaller than in Proposal 2. Proposal 1 also decreased the distance error and the standard deviation of distance error as shown in Figure 6. Therefore, filtering RSSI and the dynamic path loss index effectively reduced the variation of the positioning error. We considered that the reason why Proposal 1 did not become smaller than Proposal 2 depends on the number of vehicles shown in Section IV-F.

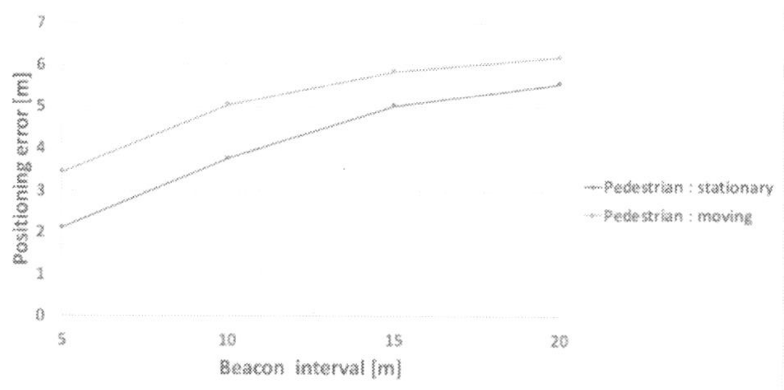

Figure 7. Relationship between beacon interval and positioning error

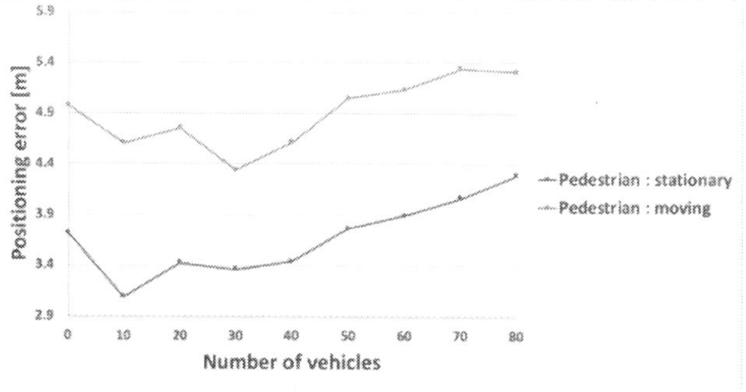

Figure 8. Relationship between number of vehicles and positioning error

E. Evaluation of beacon interval

Figure 7 shows the change in the positioning error with an increase in the beacon intervals. We found that the smaller the beacon interval is, the smaller is the positioning error. We also found that the larger the beacon interval is, the smaller is the amount of change in the positioning error.

F. Evaluation of number of vehicles

Figure 8 shows the change in positioning error with an increase in the number of vehicles. It did not become smaller as the number of vehicles increases, and it is the smallest when the number of vehicles is 10 in the stationary condition and 30 in the moving condition. This result was probably caused by using WCL for the position estimation. Because WCL calculates the weighted average of the position of vehicles and beacons, we can accurately estimate the position in a situation where the target is within a rectangle consisting of vehicles and beacons as shown in Figure 9. However, in the proposed method, we calculated positions using all the RSSIs that exceed the threshold in Section III-E. Therefore, since the available vehicle information increases when the number of vehicles

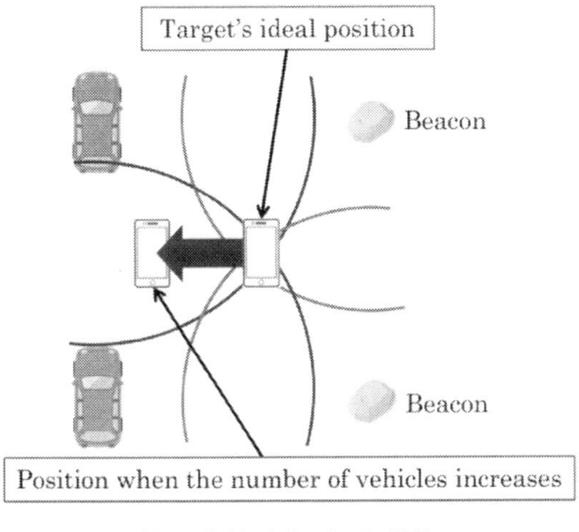

Figure 9. Ideal situation in WCL

increases, the estimated position is biased toward the vehicle side as shown in Figure 9 and that leads to poor accuracy.

V. CONCLUSION

In recent years, efforts are made for improving pedestrian safety through vehicle-to-vehicle and vehicle-to-pedestrian communication. For preventing accidents, accurate position information is critical, but GPS has low positioning accuracy in urban areas. We proposed a position estimation method using RSSI of beacons and vehicles to improve the pedestrian positioning accuracy in urban areas. When estimating positions using RSSI, the RSSI fluctuation and the static path loss index lead to distance error. In our proposed method, we applied a Kalman filter to reduce the RSSI fluctuation. By calculating the path loss index using vehicle-to-vehicle communication, we dynamically dealt with the surrounding radio wave propagation environment. Filtered RSSI and the dynamic path loss index decreased the distance error and the standard deviation of distance error. We evaluated the positioning error of the conventional methods and the proposed method by simulations and determined that our proposed method reduced the positioning error. In the future, to further improve the positioning accuracy by WCL, based on the evaluation result of the number of vehicles, we will consider a method that selects appropriate RSSI for position estimation.

ACKNOWLEDGMENT

This work was supported in part by JSPS KAKENHI Grant Number 16H02814.

REFERENCES

[1] M. Sepulcre and J. Gozalvez, "Experimental Evaluation of Cooperative Active Safety Applications Based on V2V Communications," in Proceedings of the Ninth ACM International Workshop on Vehicular Inter-networking, Systems, and Applications, ser. VANET '12. ACM, 2012, pp. 13–20, ISBN: 978-1-4503-1317-9, URL: http://doi.acm.org/10.1145/2307888.2307893 [accessed: 2018-06-03].

[2] J. J. Anaya, P. Merdrignac, O. Shagdar, F. Nashashibi, and J. E. Naranjo, "Vehicle to Pedestrian Communications for Protection of Vulnerable Road Users," in 2014 IEEE Intelligent Vehicles Symposium Proceedings. IEEE, 2014, pp. 1037–1042, ISSN: 1931-0587, URL: https://ieeexplore.ieee.org/document/6856553/ [accessed: 2018-06-03].

[3] M. Modsching, R. Kramer, and K. T. Hagen, "Field trial on GPS Accuracy in a medium size city: The influence of built-up," Proc. 3rd Workshop on Positioning, Navigation and Communication, 2006, pp. 209–218.

[4] V. Varshney, R. K. Goel, and M. A. Qadeer, "Indoor Positioning System Using Wi-Fi & Bluetooth Low Energy Technology," in 2016 Thirteenth International Conference on Wireless and Optical Communications Networks (WOCN). IEEE, 2016, pp. 1–6.

[5] Z. Jianyong, L. Haiyong, C. Zili, and L. Zhaohui, "RSSI Based Bluetooth Low Energy Indoor Positioning," in 2014 International Conference on Indoor Positioning and Indoor Navigation (IPIN). IEEE, Oct. 2014, pp. 526–533.

[6] S. Subedi, G. R. Kwon, S. Shin, S. Hwang, and J. Pyun, "Beacon Based Indoor Positioning System Using Weighted Centroid Localization Approach," in 2016 Eighth International Conference on Ubiquitous and Future Networks (ICUFN). IEEE, July 2016, pp. 1016–1019.

[7] "SPACE-TIME Engineering," URL: https://www.spacetime-eng.com/en/ [accessed : 2018-06-03].

[8] "ITU-R Recommendation P.1411-9," URL: https://www.itu.int/dms_pubrec/itu-r/rec/p/R-REC-P.1411-9-201706-I!!PDF-E.pdf [accessed : 2018-06-03].

Improvement of False Positives in Misbehavior Detection

Shuntaro Azuma
Computer and Information Science
Graduate School of Science and Engineering
Doshisha University
Kyoto, Japan
email:syuntaro.azuma@nislab.doshisha.ac.jp

Manabu Tsukada
Graduate School of Information
Science and Technology
Tokyo University
Tokyo, Japan
email:tsukada@hongo.wide.ad.jp

Kenya Sato
Computer and Information Science
Graduate School of Science and Engineering
Doshisha University
Kyoto, Japan
email:ksato@mail.doshisha.ac.jp

***Abstract*—By faking vehicle information on cloud servers, an adversary may deliberately cause traffic congestion and/or accidents. Misbehavior means sending masqueraded data to cloud servers in this paper. In our previous research, we proposed "A Method of Detecting Camouflage Data with Mutual Position Monitoring". Cloud servers can detect masqueraded position data from malicious vehicles by increasing the threshold value of our detecting method. However, there are some problems. In this paper, we clarify what kind of malicious behavior is targeted, and we propose two new measures to address the false positives problem. First, we weight for public vehicles such as police cars, and cloud servers can trust vehicles even if they below the threshold value. Second, we dynamically determine the threshold value with consideration of vehicle density. Next, we evaluate the two methods. We find that the method of weighting for each vehicle was very effective, and the method of dynamic determination also showed good results. There is not much difference between our previous method and weighting for each vehicle at low threshold value, but this new one helps considerably suppress false positives at high threshold. The advantage of the dynamic determination model is that false positives do not depend on each base station, because the threshold is dynamically determined. This works more effectively in lower vehicle densities. Our results indicated that these two countermeasures was practical against false positives.**

Keywords–vehicle security; V2X communication; misbehavior detection.

I. INTRODUCTION

In recent years, research on autonomous driving and vehicle-to-vehicle (V2V) communication have been conducted in the Intelligent Transport Systems (ITS) field. In addition, vehicles have vehicle-to-cloud (V2C) communication with cloud servers using mobile lines. When vehicles are connected to various targets, malicious acts have enormous impact. This paper represents further work on our previous publication "A Method of Detecting Camouflage Data with Mutual Position Monitoring" [1]. In our previous research, we proposed how to detect malicious vehicles which sent masqueraded data of their positions. We evaluated the detection rates and received good results. We found that we could detect completely malicious vehicles by increasing the threshold value of our detecting method. However, we have some problems. We especially considered the false positives problem in our previous research. We thought that vehicle densities affect false positives, so we calculated them in high vehicle densities. We cloud find high vehicle densities help suppress false positives, but this countermeasure is effective in only this situation. We should address the false positives problem in low vehicle densities. In this paper, we will reveal our research's target at first. Next, we will describe the operation of proposed method. Then, we will describe improvements of previous research, which are methods of weighting for each vehicle and dynamic determination, and then we will describe the evaluation of these methods.

II. THREAT ANALYSIS OF TRANSMISSION DATA

There exits previous works researching the detection of malicious vehicles in V2X communication [2] [3], as a matter of fact, the definition of a malicious vehicle is ambiguous. In this section, we analyze attacks on vehicle communication and clarify what kind of malicious vehicles are

A. Threat Analysis of Transmission Data

Table I shows the threat analysis of data transmitted to a cloud server. These threats include eavesdropping attacks, falsifications, and spoofing. Spoofing attacks are divided into vehicle impersonation and data masquerade. Vehicle impersonation means that attackers pretend to be other vehicles. For example, even though one vehicle does not have any trouble, an attacker pretends to be another vehicle and then calls the police lying that it had an accident. An example of data masquerade is when a vehicle's own position information or status is masked.

Security requirements regarding these threats include confidentiality, completeness, node reliability, and data reliability. To supply confidentiality and completeness, data encryption is proposed and can be done by a secret key or an ID base cipher. Node reliability identifies vehicles that are pretending to be other vehicles. The Public Key Infrastructure (PKI) method, which is adapted by the vehicles, is one good resolution because certificates guarantee vehicles. Data reliability prevents attackers from masquerading data. However, this is not effective for all spoofing acts.

TABLE I. THREATS ANALYSIS ABOUT TRANSMISSION DATA

THREAT		REQUIREMENT	COUNTERMEASURE
Eavesdropping		Confidentiality	Encryption
Falsification		Completeness	Encryption
Spoofing	Vehicle impersonation	Node reliability	PKI
	Data masquerade	Date reliability	**Target of this research**

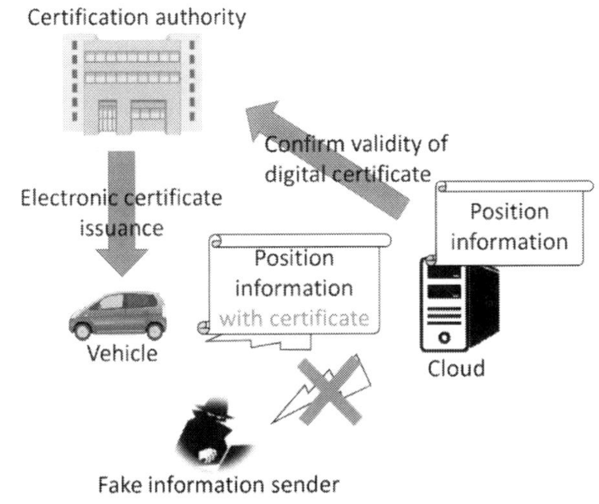

Figure 1. PKI to adapt to vehicles

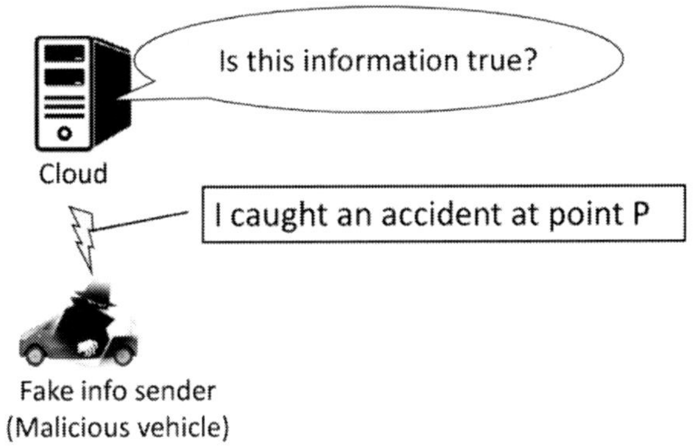

Figure 2. Problem of settling by this research

B. *Difference Between Node and Data Reliability*

Node reliability means that a cloud server trusts a particular vehicle and believes that it is not pretending to be a different vehicle. The previous section showed that the PKI method can be adapted by vehicles to resolve this problem. A cloud may be able to verify the electronic certification and confirm the transmitter's information by the mechanism shown in Figure 1.

However, this research focuses on data masquerade, as described in Figure 2. Since data encryption and PKI do not confirm whether the received data are masqueraded, data masquerade is inherently different from node reliability which can be resolved by these methods. We will propose a method that can handle such example, which guarantees the reliability of the data.

III. OUR PREVIOUS RESEARCH

In this section, we will explain our previous research again. We use vehicle-to-everything (V2X) communication and detect masqueraded data of vehicle's position.

A. *Pre-suppositions*

1) A safe channel has been secured by relationships of mutual trust among all vehicles and cloud servers.

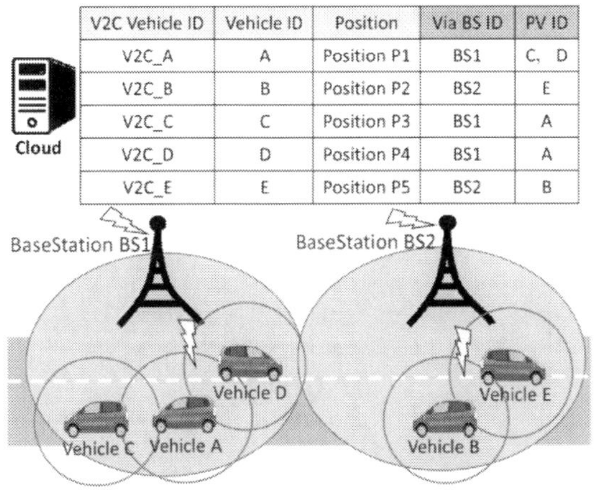

Figure 3. Use example of peripheral vehicle information in V2X communication

2) Vehicles and cloud servers have been mutually certified beforehand.
3) Relationships between cloud servers and base stations have been built.

B. *Definition of Terminology in Proposed Method*

- Vehicle ID

This ID is used by vehicles in V2V communication, and this is a different public ID for each vehicle.

- V2C Vehicle ID

This ID is used for a unique key in V2C communication. This secret ID is not available to others. V2C Vehicle ID and Vehicle ID is uniquely related.

- Via Base Station (Via BS) ID

This ID is used in V2C communication, and this is a different ID for each base station.

- Peripheral Vehicle (PV) ID

This ID is a received vehicle ID from other vehicles in V2V communication.

C. *Outline*

Vehicles can use V2X communication. When they send their position information to a cloud server, they also send other information in addition to their position. In this research, a cloud server detects masqueraded data from transmitted data by using the relay base station information in vehicle-to-cloud (V2C) communication and peripheral vehicles in vehicle-to-vehicle (V2V) communication.

Figure 3 shows the picture of misbehavior detection in our previous research, and Figure 4 shows how to detect masqueraded data in a cloud. A cloud receives not only position information or VehicleID but also peripheral vehicle's and relay base station's information.

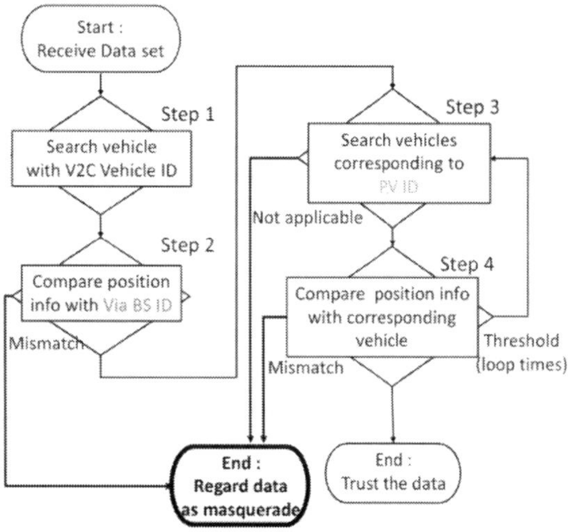

Figure 4. Misbehavior data detection procedure

D. How to Detect Misbehavior data

V2CVehicleID is used in the first step on Figure 4. Cloud servers confirm whether received data is sent from vehicles or not. Second, cloud servers compare Via Base Station ID (ViaBSID) with received position information to confirm whether a sending vehicle exists in relay base station's coverage area. When the received position information exceeds this area, we assume that it can't be consistent and that received information was regarded as masqueraded data. This step helps detect data masquerade toward other base station's coverage area. At the third and fourth step, cloud servers detect masqueraded data by using peripheral vehicle IDs (PVIDs). Third, cloud servers search vehicles corresponding to sending vehicle's PVID. Firth, cloud servers compare the received position with peripheral vehicle's position corresponding to PVID. If the distance between two vehicle's position exceeds V2V communication coverage, we assume that it can't be consistent and that received information was regarded as masqueraded data. This operation is performed a predetermined number of times. In the proposed method, a predetermined number of times means the number of PVIDs which is necessary for cloud servers to trust. This is a so-called threshold value. By setting this threshold, we can assure more reliable data.

E. Advantage of This Proposal

Figure 5 shows a countermeasure example of position data masquerade. We can detect masqueraded position information toward another base station using relay base station's information in V2C communication. In addition, Figure 6 shows a countermeasure example of position data masquerade. We assume that a malicious vehicle masquerades its own position information. A cloud confirms PVIDs sent from a vehicle and compares received position information with peripheral vehicle's positions which are relevant to PVIDs. When a cloud finds that transmitted position information is outside V2V communication coverage with peripheral vehicles, the cloud determines that the received position information has been masqueraded. However when this information does not exceed the coverage area, the cloud trusts the received position information. Vehicles acquire peripheral vehicle information in V2V communication and mutually monitor them. This helps cloud servers detect masqueraded data.

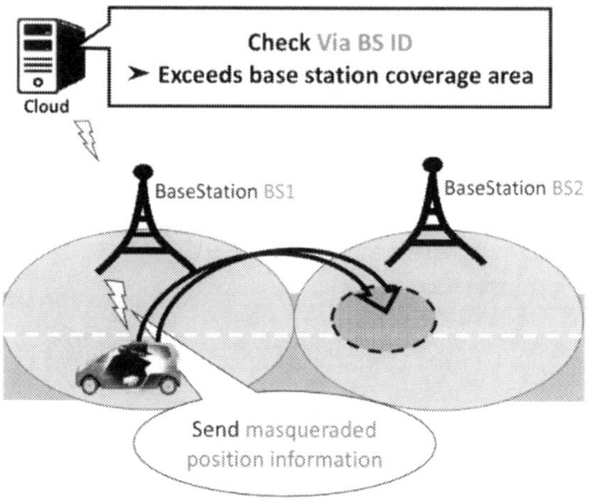

Figure 5. Advantage of using base station information

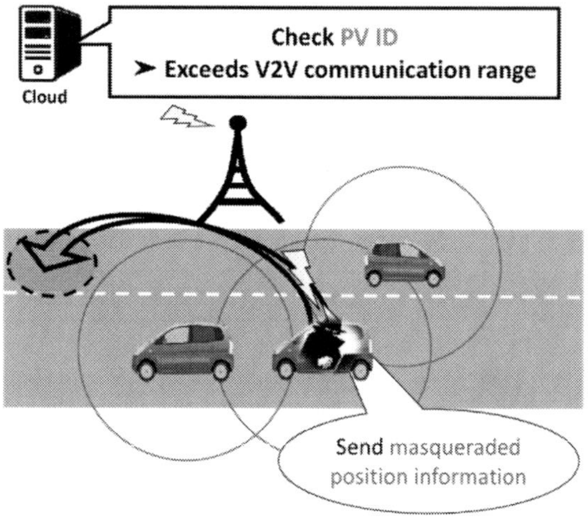

Figure 6. Advantage of using peripheral vehicle information

IV. DEVELOPMENT OF OUR PROPOSED METHOD

We have some problems, especially false positives. Therefore, we propose here two new points to solve them.

1) We weight the public vehicles and trust cloud servers more even for vehicles below the threshold.
2) We dynamically determine the threshold value with consideration of vehicle density.

A. False Positives

We know that increasing the threshold in our method can help detect masqueraded data. However, when we increased

TABLE II. PREVIOUS SIMULATION PARAMETER

Simulator	Scenargie2.0	
Vehicle number	158 [cars] (five of the send masquerade positions.)	
Area	1000 [m] × 1000 [m]	
Communication mode	ARIB STD T109	LTE
Use frequency band	700 [Mhz]	2.5 [GHz]
Communication interval	100 [ms]	1.0 [s]
Radio spread model	ITU-R P.1411	LTE-Macro
Base station ground clearance	1.5 [m]	

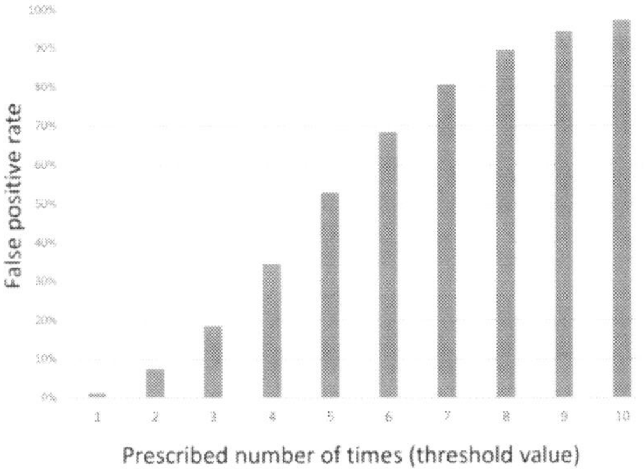

Figure 7. False positives by threshold value under Japanese average vehicle density (158[cars/km^2]) environment

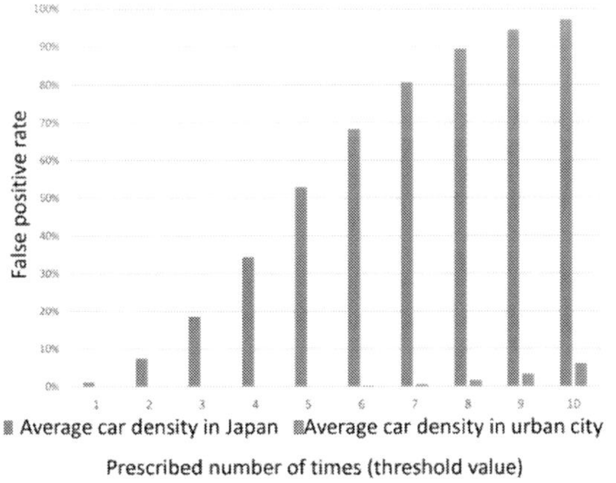

Figure 8. False positives comparison with urban area average vehicle density (1128 [cars/km^2]) environment

threshold values, false positive rates dramatically increased. Therefore, we considered the false positives problem in our previous research. Figure 7 shows false detection rates (false positives) of our proposed method, which is based on the average vehicle density in Japan. The method's threshold is the number of PVIDs, which is necessary for cloud servers to trust. In the previous simulation environment shown in Table II, Figure 7 shows the false positives when all 158 cars are not misbehaving. By increasing the threshold value, false positive rates increased. By increasing the threshold value under Japanese average vehicle density, cloud servers erroneously detect normal communication as abnormal.

Then, the false positive rates under the average vehicle density environment in urban city (Osaka), which has the highest average car density in Japan, are shown in Figure 8. In a high vehicle density area, since vehicles can acquire a lot of peripheral vehicle information in V2V communication, even if the threshold is increased, an increase of the false detection rate can be suppressed. Tables III and IV show precision, recall, and F-measure in our proposed method. Even looking at these tables, we can make the same statement as above.

TABLE III. F-MEASURE UNDER 158[cars/km^2] ENVIRONMENT

Threshold	1	2	3	4	5	6	7	8	9	10
Precision	1.0	1.0	1.0	1.0	1.0	1.0	1.0	1.0	1.0	1.0
Recall	0.99	0.93	0.81	0.66	0.47	0.32	0.19	0.11	0.056	0.029
F-measure	0.99	0.96	0.90	0.79	0.64	0.48	0.32	0.19	0.11	0.056

TABLE IV. F-MEASURE UNDER 1128[cars/km^2] ENVIRONMENT

Threshold	1	2	3	4	5	6	7	8	9	10
Precision	1.0	1.0	1.0	1.0	1.0	1.0	1.0	1.0	1.0	1.0
Recall	1.00	1.00	1.00	1.00	1.00	1.00	0.99	0.98	0.97	0.94
F-measure	1.00	1.00	1.00	1.00	1.00	1.00	1.00	0.99	0.98	0.97

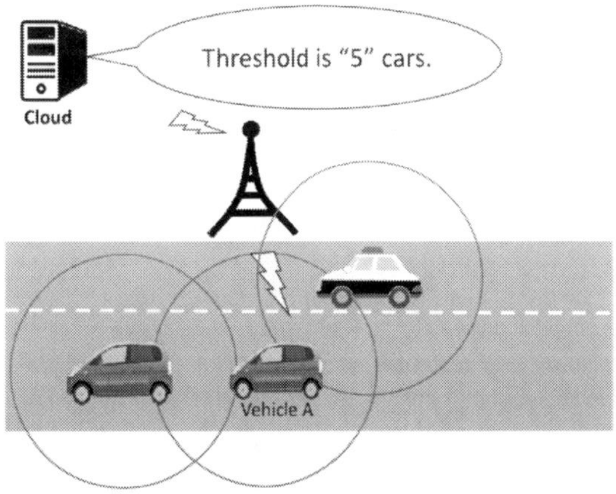

Figure 9. Examples when using a method of weighting for each vehicle

B. Weighting for Each Vehicle

This good result (Figure 8) only applies in the urban area. We need to take another measure under the environment of Japanese average car density. In addition, we must consider the lesser number of cars in the streets at nighttime and the lower density environment. Figure 10 shows our new countermeasure to the false positives. We give more weight to public vehicles such as police vehicles and buses than normal vehicles. Even if the vehicle communicating with the public vehicle (that is, the vehicle including the public vehicle in peripheral vehicle

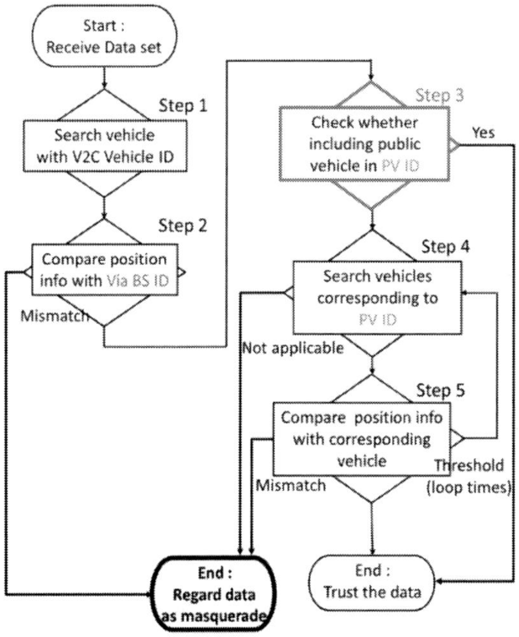

Figure 10. New flowchart of weighting for each vehicle

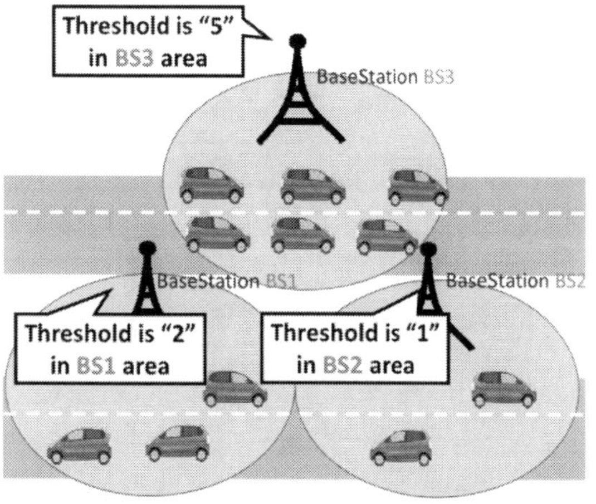

Figure 11. Examples when using dynamically threshold determination

information) does not exceed the threshold value, this one is trusted by a cloud. We consider the environment shown in Figure 9. This case is that the threshold required for the cloud to trust is 5. Vehicle A has only three peripheral vehicles. But because there are a police vehicle in them, a cloud trusts vehicle A. We think that this method will reduce the false positives if public vehicles are running even in low vehicle density areas.

C. Dynamic Determination of Threshold

Based on the results (shown in Figures 7 and 8), we calculate vehicle density for each base station and change the threshold value for each base station. Figure 11 shows the overall picture. We set a prescribed percentage as the threshold value. When vehicle densities in base stations change, the threshold also changes for each base station. We think that this method is effective in solving the false positives problem because we can adjust the threshold dynamically in areas where a vehicle density is low, or during times when there are few vehicles.

V. EVALUATION AND CONSIDERATION

We will calculate false positive rates to evaluate our new points. Next, we will consider the practicality of our new points from the evaluation obtained.

A. Simulator

In this paper, we use Scenargie [4] as a simulator to evaluate the performance of our proposed method. Scenargie is a network simulator developed by Space-Time Engineering (STE). By combining expansion modules, such as LTE, V2V communication and multi-agent, we can construct a realistic simulation. In addition, since communication systems and evaluation scenarios are becoming more complicated, this ingenious simulation has greatly reduced the effort required to create scenarios.

B. Evaluation Model

For an evaluation environment, we use one square kilometer Manhattan model and use simulation parameters shown in Table V. We set the number of vehicles to 158 [cars]

TABLE V. SIMULATION PARAMETER

Simulator	Scenargie2.0	
Vehicle number	158 [cars] (including two police cars.)	
Area	1000 [m] × 1000 [m]	
Communication mode	ARIB STD T109	LTE
Use frequency band	700 [Mhz]	2.5 [GHz]
Communication interval	100 [ms]	1.0 [s]
Radio spread model	ITU-R P.1411	LTE-Macro
Base station ground clearance	1.5 [m]	

and the range to 1 [km^2] because the average car density in Japan is 158 [cars/km^2]. ITU-R P.1411 model is a radio wave propagation scheme that considers road map information, and radio waves are attenuated based on the shape of the road, so we compared with a two-ray model, which includes direct waves and reflected waves from the ground, this model is close to reality. ITU-R P.1411 model is a radio wave propagation scheme that considers road map information, and radio waves are attenuated based on the shape of the road, so we compared with a two-ray model, which includes direct waves and reflected waves from the ground, this model is close to reality.

C. Evaluation of Weighting for Each Vehicle

Figure 12 shows false positive rates when using a method of weighting for each vehicle, and Table VI shows precision, recall, and F-measure. Comparing to Figure 7, we can find that false positives are considerably suppressed at high threshold values. When the threshold is low, we do not find much difference. Therefore, we say that public vehicles have little influence on false positives at low threshold values. However, when a vehicle communicates with a public vehicle in this method, cloud servers can trust this one even if its own PVID has not reached the threshold value. Even if there are no peripheral vehicles around vehicles which send their position data to cloud server, but public vehicles driving around them,

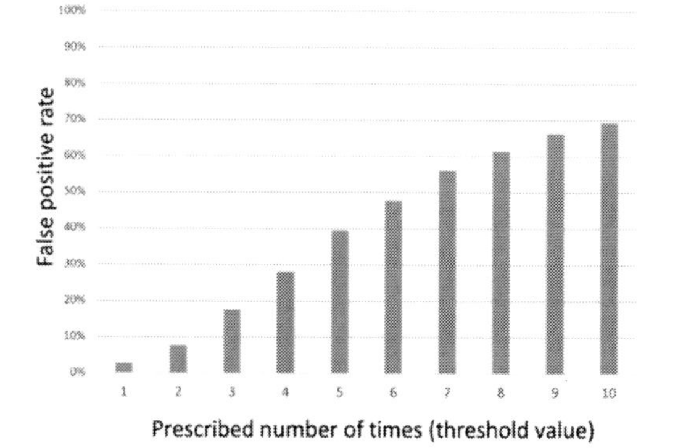

Figure 12. False positives by threshold value when using a method of weighting for each vehicle

TABLE VI. F-MEASURE WITH A METHOD OF WEIGHTING FOR EACH VEHICLE

Threshold	1	2	3	4	5	6	7	8	9	10
Precision	1.0	1.0	1.0	1.0	1.0	1.0	1.0	1.0	1.0	1.0
Recall	0.97	0.92	0.83	0.72	0.61	0.52	0.44	0.39	0.34	0.31
F-measure	0.99	0.96	0.90	0.84	0.75	0.69	0.61	0.56	0.51	0.47

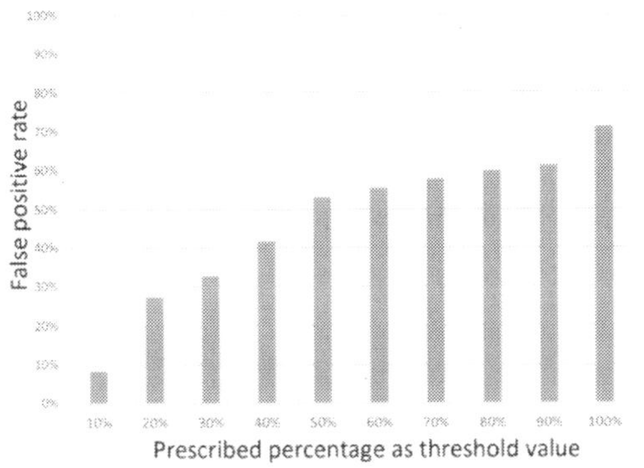

Figure 13. False positives by prescribed percentage when using dynamic threshold determination

TABLE VII. F-MEASURE WITH A METHOD OF DYNAMIC THRESHOLD DETERMINATION

Threshold	1	2	3	4	5	6	7	8	9	10
Precision	1.0	1.0	1.0	1.0	1.0	1.0	1.0	1.0	1.0	1.0
Recall	0.92	0.73	0.67	0.58	0.47	0.45	0.42	0.40	0.39	0.29
F-measure	0.96	0.84	0.80	0.74	0.64	0.62	0.59	0.57	0.56	0.45

they can be trusted by cloud servers. Our proposed method works more effectively at high threshold. We focused on police cars as public vehicles in this paper, but we guess that we further suppress false positives by weighting buses or taxis running throughout the city.

D. *Evaluation of Dynamically Threshold Determination*

Cloud servers dynamically determine threshold values at each base station by confirming vehicle densities in base station's coverage area. We calculate false positive rates with this method. Figure 13 shows false positives when using dynamic determination of threshold. This graph's horizontal axis is the ratio of base station's vehicle density as the threshold. It means that cloud servers calculate the number of vehicles traveling in the base station, and we consider the predetermined percentage as the threshold value. Therefore, we do not know the accurate threshold value because there are different vehicle densities for each base station. At low percentage of Figure 13, because the threshold value is lower in each base station, we can suppress false positive rates. For example, when there are 30 vehicles in a base station's coverage area and prescribed percentage is 10%, the threshold value becomes 3. Therefore, cloud servers trust vehicles which have three PVIDs. However, when we set 100% as prescribed percentage in 30 driving vehicles environment, the threshold value becomes 30, so vehicles should communicate with other thirty vehicles for cloud servers trusting them. As the percentage increases, the threshold increases, therefore false positives increase. The advantage of this method is that false positives do not depend on each base station, because the threshold is dynamically determined. If we decide on a single threshold, cloud servers will not respond flexibly.

VI. CONCLUSION

In the Intelligent Transport Systems (ITS), using cloud servers is inevitable. In our previous research, we used V2X communication, obtained information from various objects, and described measures against data masquerade. We cloud completely detect masqueraded data by increasing threshold values. However, we have some problems, which are false positives especially. We proposed two countermeasures against the false positives problem. First, we weight the public vehicles and trust more on cloud servers even for vehicles below the threshold. Second, we dynamically determine the threshold value with consideration of vehicle density. As a result of evaluating these, we succeeded in suppressing false positives. In particular, the method of weighting each vehicle has proven more effective. There is not much difference between previous results and this paper' s results at low threshold values, but at high threshold values, this method help suppress false positive rates. The second measure means that could servers calculate the number of vehicles traveling in the base station, and we consider the predetermined percentage as the threshold value. The advantage of this method is that false positives do not depend on each base station, because the threshold is dynamically determined. In this research, we think that we have improved considerably the false positives problem. In the future, we will propose a method combining both methods or a completely new method, and we would like to conduct a demonstration experiment that also cooperates with Local Dynamic Map (LDM).

ACKNOWLEDGMENT

A part of this work was supported by KAKENHI (JP 16H02814) and MEXT program for the strategic research foundation at private universities.

REFERENCES

[1] Shuntaro Azuma, Manabu Tsuakda, and Kenya Sato" A Method of Detecting Camouflage Data with Mutual Vehicle Position Monitoring," VEHICULAR 2017, July 2017.

[2] Yang, Yuchen Ou, Dongxiu Xue, Lixia Dong, and Decun" Infrastructure-based Detection Scheme of Malicious Vehicles for Urban Vehicular Network," Transportation Research Board 96th Annual Meeting, January 2017.

[3] Gongjun Yan, Stephan Olariu, and Michele C. Weigle" Providing VANET security through active position detection," Computer Communications, vol. 31, pp. 2883-2897, July 2008.

[4] SPACE-TIME Engineering. Available from: https://www.spacetime-eng.com/en/ 2017.07.07